INTERACTIVE EXPLORATIONS
CD-ROM FOR MACINTOSH® AND WINDOWS®
TEACHER'S GUIDE

HOLT, RINEHART AND WINSTON
A Harcourt Education Company

Orlando • **Austin** • New York • San Diego • Toronto • London

To the Teacher

Imagine having access to a fully equipped laboratory where your students could study questions and problems related to life processes, geology, energy and power, and chemical reactions. This is exactly what is possible with *Holt Science and Technology Interactive Explorations*. By using this innovative CD-ROM program, students get valuable experience in a unique laboratory setting.

Welcome to Dr. Crystal Labcoat's laboratory, where with the click of a mouse, students will have access to a variety of scientific tools and equipment and will be challenged to solve some perplexing problems and mysteries. Dr. Labcoat operates a virtual laboratory, and your students are her lab assistants. Under her guidance, your students will perform some amazing and highly interesting scientific experiments and studies. But be prepared—although Dr. Labcoat provides the lab and the equipment, your students provide the brainpower.

This *Teacher's Guide* consists of the following components:

- **User's Guide**

  The User's Guide provides important technical information about the program, including its installation, features, and use.

- **Teaching Notes, Worksheets, and Handouts**

  Organized by Exploration, this information includes background material and worksheets that guide students through the CD-ROM experience and allow them to record their answers on paper rather than electronically. In addition, the Computer Database articles (also called CD-ROM articles) are provided so that they can be used as handouts. The worksheets and handouts make the program more flexible in cooperative groups or when computer time is limited.

- **Answer Keys**

  Worksheet pages with overprinted answers are provided for each Exploration to make grading worksheets and fax forms fast and efficient.

Copyright © by Holt, Rinehart and Winston

All rights reserved. No part of this publication may be reproduced or transmitted in any form or by any means, electronic or mechanical, including photocopy, recording, or any information storage and retrieval system, without permission in writing from the publisher.

Teachers using HOLT SCIENCE AND TECHNOLOGY may photocopy blackline masters in complete pages in sufficient quantities for classroom use only and not for resale.

Art/Photo Credits
All work contributed by Holt, Rinehart and Winston except the following:
Front Cover (zebra), JH Pete Carmichael/Getty Images; (arch), Steve Niedirf Photography/Getty Images; (aircraft), Creatas/PictureQuest; (owl), Kim Taylor/Bruce Coleman
Disc 3, Exploration 2, USS *Birmingham* submarine courtesy of U.S. Naval Institute; Disc 3, Exploration 4, Antarctica map from Mountain High Maps® Copyright © 1995 Digital Wisdom, Inc.

Printed in the United States of America

0-03-035691-1   3 4 5 6 7 8 9   085   09 08 07 06 05 04

Contents

Disc 1 Contents iii
Disc 2 Contents vi
Disc 3 Contents ix
Contents by Subject xii
User's Guide xiii
    System Requirements xiii
    Installing the Program xiv
    Logging On xv
    Welcome to the Main Menu xvi
    Using the Virtual Laboratory xvii
    Assessment Tools xx
    Networking Student Reports xxii
    Optimizing Performance xxiii
    Technical Support Information xxiv
Disc 1 Answer Key 231
Disc 2 Answer Key 247
Disc 3 Answer Key 263

Disc 1 Contents

Exploration 1: Something's Fishy 1

African Cichlids are dying at the local fish store. The store's manager, Ray McMullet, needs to know what changes to make in order to save the fish and prevent this trauma at the Fishorama.

LIFE SCIENCE

Disc 1 Contents, continued

Exploration 2: Shut Your Trap! 9

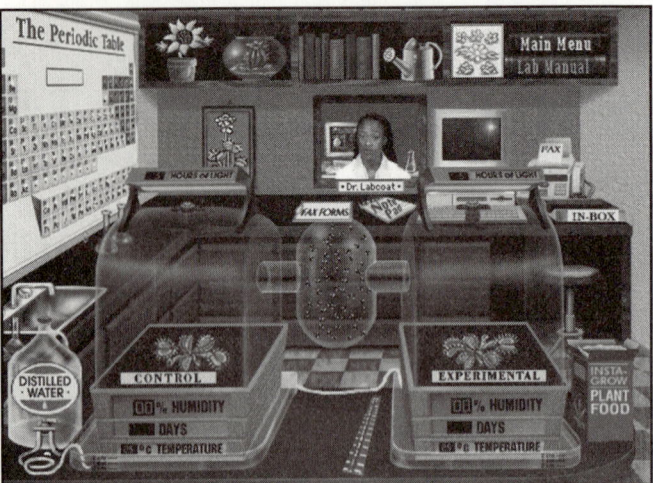

Plant poachers are threatening the existence of Venus' flytraps in the wild. Lily N. Lotus wants to find out what the optimal growth conditions are for these carnivorous plants so that they may be easily grown in nurseries.

LIFE SCIENCE

Exploration 3: Scope It Out! 18

Some microorganisms have been extracted from the digestive tract of an ancient, amber-entombed bee. Dr. Viola Russ wants to classify these gutsy microorganisms and explain the role that they might have played in the bee's life.

LIFE SCIENCE

Exploration 4: What's the Matter? 27

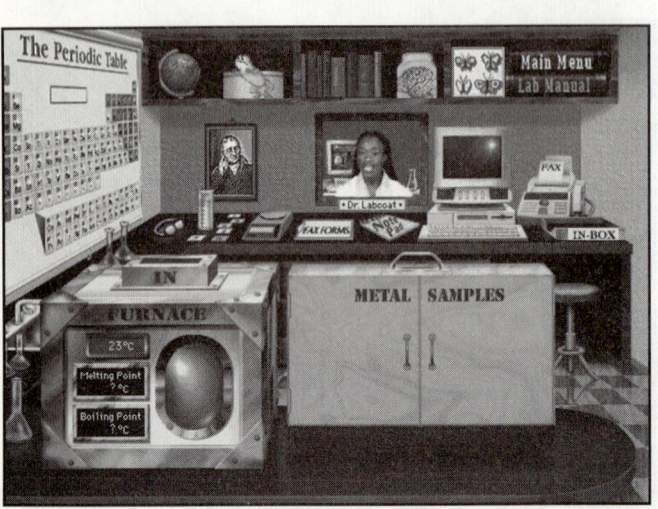

Scientists in Hawaii are in a hot spot—the tip of one of their instruments, the lava analyzer, has melted. Dr. John Stokes and his team need to know the most practical metal to use to replace the tip of the lava analyzer.

INTEGRATED SCIENCE
PHYSICAL/EARTH

Disc 1 Contents, continued

Exploration 5: Element of Surprise 38

Fred Stamp is packing some samples of chemical elements and shipping them to the South Pole. He needs to ensure that the shipment reaches a remote research station safely without encountering any explosive surprises.

PHYSICAL SCIENCE

Exploration 6: The Generation Gap 48

Wendy Powers operates a small log-home manufacturing plant called EcoCabin, Inc. She wants to know if the Electroprop, a wind turbine, would help her customers generate some savings on their electricity bills.

EARTH SCIENCE

Exploration 7: Teach It While It's Hot! 58

Mr. McCool is teaching a lesson on the relationship between temperature and heat to his class of middle-school students. He wants to make sure that this lesson is cool enough to get his students fired up about temperature and heat.

PHYSICAL SCIENCE

Disc 1 Contents, continued

Exploration 8: Flood Bank 68

A local environmental-impact committee is debating a reservoir of pros and cons about whether to build a dam on a nearby river. The committee chairperson, Sandy Banks, wants to find out what impact the dam would have on the local river environment.

INTEGRATED SCIENCE
LIFE/EARTH

Disc 1 Answer Key 231

Disc 2 Contents

Exploration 1: What's Bugging You? 77

Several of Dr. Mike Roe's patients are suffering from a baffling illness. Dr. Roe suspects it is caused by one of several organisms, but he needs help figuring out exactly which organism is bugging his patients.

LIFE SCIENCE

Exploration 2: Sea Sick 87

Some unidentified sea creatures have been delivered to Shelley C. Waters at the Marine Exploratorium. Ms. Waters is deeply concerned about their sickly appearance and wants to know what they are and how to make them thrive.

INTEGRATED SCIENCE
LIFE/EARTH

Disc 2 Contents, continued

Exploration 3: Moose Malady 97

Moose at a preserve in western Sweden are dying. The preserve's director, Hans Oleson, needs to determine the root cause of this mysterious malady so that he can help save the remaining moose.

INTEGRATED SCIENCE
LIFE/EARTH

Exploration 4: Force in the Forest 106

Gustavo Solimões is the driving force behind a small manufacturing facility that exports native products from a Brazilian rain forest. He needs some advice about how to use a pushing device to set his production line in motion.

PHYSICAL SCIENCE

Exploration 5: Extreme Skiing 114

A manufacturing company is designing an artificial leg for a ski racer who plans to compete in the next Paralympic Games. The company's director, Ludwig Guttman, needs help selecting the best material to use for this unusual application.

PHYSICAL SCIENCE

CONTENTS vii

Disc 2 Contents, continued

Exploration 6: Rock On! 123

Claudia Stone is constructing an information kiosk for a new state park. She wants to know the identity and classification of several rock specimens from the area so that the kiosk will provide some rockin' geological history.

INTEGRATED SCIENCE
LIFE/EARTH

Exploration 7: Space Case 133

An unfamiliar piece of equipment has been donated to Desert View Park. Estelle de la Luna, the park's director, needs to figure out how to use this equipment to educate the park's many star-struck visitors about astronomy.

EARTH SCIENCE

Exploration 8: How's It Growing? 143

Rosie Flores is in a colorful situation—she is filling in for a gardening expert who writes a question-and-answer column in the newspaper. Ms. Flores needs help responding to the growing concerns of a hydrangea enthusiast.

LIFE SCIENCE

Disc 2 Answer Key 247

Disc 3 Contents

Exploration 1: The Nose Knows 153

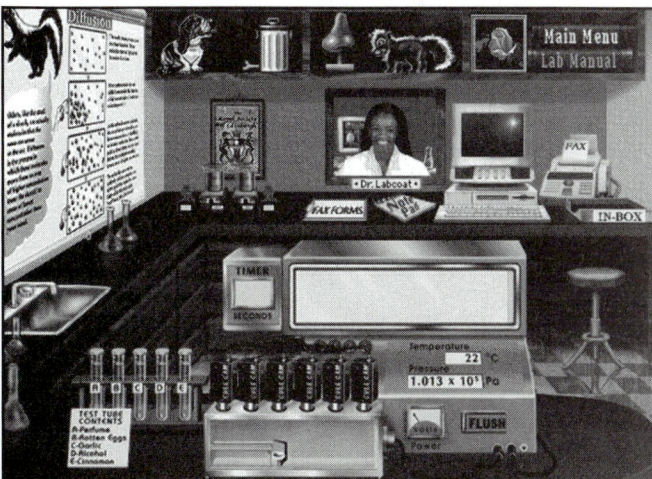

Ms. Dee Foushen needs help sniffing out the solution to a problem. She wants to know the best chemical to use for an odor alarm that would warn her hearing- and sight-impaired students in the event of a fire.

PHYSICAL SCIENCE

Exploration 2: Sea the Light 161

Ms. Diane Sittie has come up with a bright idea: an underwater lamp for scuba divers that neither sinks nor floats. Unfortunately, she is in the dark about some of the details of her invention and needs help keeping her idea afloat.

PHYSICAL SCIENCE

Exploration 3: Stranger Than Friction 171

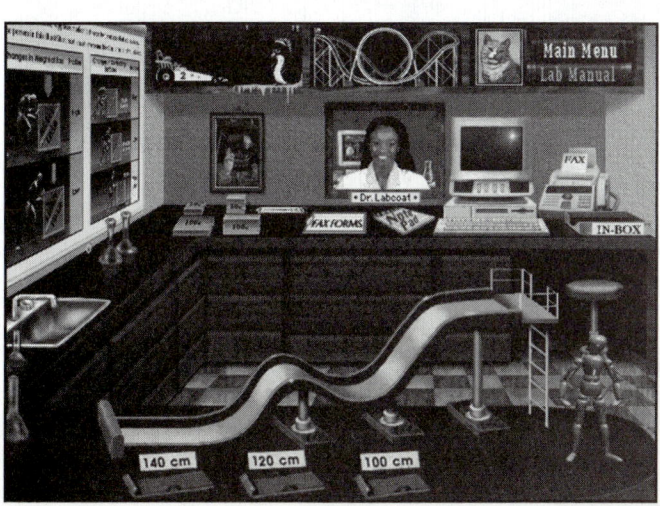

Mr. Norm N. Cline is fighting an uphill battle in the design of an amusement park ride that consists of a slide and toboggans. He needs help selecting the best materials and toboggan size to avoid unnecessary friction with park patrons.

PHYSICAL SCIENCE

Disc 3 Contents, continued

Exploration 4: Latitude Attitude 180

Research scientists on the Mertz Glacier in Antarctica are expecting an air-delivery of fresh supplies. Captain Corey O. Lease must confirm her flight plan so that she can straighten up and fly right to the research site.

EARTH SCIENCE

Exploration 5: Tunnel Vision 188

Seymore Rhodes wants to create a lighted bicycle helmet that will make riding his bike to school safer. Unfortunately, he's not an electrical genius—he needs help seeing the light at the end of the tunnel.

PHYSICAL SCIENCE

Exploration 6: Sound Bite! 197

A loud humming noise from a new ice cream shop has some guinea pigs at the neighboring pet store in an uproar. Mr. Cy Lintz wants to know how to use active sound control to give his guinea pigs some peace and quiet.

PHYSICAL SCIENCE

Disc 3 Contents, continued

Exploration 7: In the Spotlight 207

Ms. Iris Kones' first production at a community theater is really in the spotlight. She needs help filtering through her lighting options to determine how many different colors of light she can produce.

PHYSICAL SCIENCE

Exploration 8: DNA Pawprints 218

Ms. Jean Poole, a local dog breeder, is afraid she's barking up the wrong tree when it comes to completing pedigrees. She needs help figuring out which of her male dogs sired her younger dogs so that she can enter the pups in a dog show.

LIFE SCIENCE

Disc 3 Answer Key 263

Answer Key 230

Contents by Subject

LIFE SCIENCE

Title	Location
Something's Fishy	Disc 1 Exploration 1
Shut Your Trap!	Disc 1 Exploration 2
Scope It Out!	Disc 1 Exploration 3
Flood Bank	Disc 1 Exploration 8
What's Bugging You?	Disc 2 Exploration 1
Sea Sick	Disc 2 Exploration 2
Moose Malady	Disc 2 Exploration 3
Rock On!	Disc 2 Exploration 6
How's It Growing?	Disc 2 Exploration 8
DNA Pawprints	Disc 3 Exploration 8

EARTH SCIENCE

Title	Location
What's the Matter?	Disc 1 Exploration 4
The Generation Gap	Disc 1 Exploration 6
Flood Bank	Disc 1 Exploration 8
Sea Sick	Disc 2 Exploration 2
Moose Malady	Disc 2 Exploration 3
Rock On!	Disc 2 Exploration 6
Space Case	Disc 2 Exploration 7
Latitude Attitude	Disc 3 Exploration 4

PHYSICAL SCIENCE

Title	Location
What's the Matter?	Disc 1 Exploration 4
Element of Surprise	Disc 1 Exploration 5
The Generation Gap	Disc 1 Exploration 6
Teach It While It's Hot!	Disc 1 Exploration 7
Force in the Forest	Disc 2 Exploration 4
Extreme Skiing	Disc 2 Exploration 5
The Nose Knows	Disc 3 Exploration 1
Sea the Light	Disc 3 Exploration 2
Stranger Than Friction	Disc 3 Exploration 3
Tunnel Vision	Disc 3 Exploration 5
Sound Bite!	Disc 3 Exploration 6
In the Spotlight	Disc 3 Exploration 7

INTEGRATED SCIENCE

Title	Location
What's the Matter?	Disc 1 Exploration 4
Flood Bank	Disc 1 Exploration 8
Sea Sick	Disc 2 Exploration 2
Moose Malady	Disc 2 Exploration 3
Rock On!	Disc 2 Exploration 6

User's Guide

SYSTEM REQUIREMENTS

Before you begin using *Holt Science and Technology Interactive Explorations,* you will need to acquire the necessary equipment and set it up properly. The complete setup includes a computer (IBM®-compatible or Macintosh®-compatible) connected to a CD-ROM drive. Audio headphones are optional. To use the program with a network, you will also need additional cables to connect the machines, a dedicated network file server, and a network operating system.

A note concerning minimum requirements: *Although the program will run on machines with the minimum requirements listed below, we strongly recommend that the program be used on newer model computers (040 Macintoshes and 486 PCs or higher) that are more efficient at handling the demands of multimedia. If you run the program on lower-end machines, you may experience slow response times to mouse clicks, longer loading times for video and sound, and slower animations, as well as occasional dropped video frames and sound.*

COMPUTERS

Macintosh-Compatible Computers

Minimum Requirements:
- 68030 CPU running at 25 MHz or higher **(Highly recommended: 68040 CPU)**
- Color display capable of 256 colors at 640 × 480 resolution
- System 7.1 or higher
- Double-speed or higher CD-ROM drive
- 5 MB of RAM available for application
- 30–40 MB of free memory on hard drive if you plan to install individual Explorations (Explorations can also be run directly from the CD-ROM)
- Internal/external speaker(s); headphones (recommended for classroom settings)

IBM-Compatible Computers

Minimum Requirements:
- 80386 DX running at 25 MHz or higher **(Highly recommended: 80486 CPU)**
- Color display capable of displaying 256 colors at 640 × 480 resolution
- Windows® 3.1, Windows 95, or Windows 98
- Double-speed or higher CD-ROM drive
- 8 MB of RAM
- 30–40 MB of free memory on hard drive if you plan to install individual Explorations (Explorations can also be run directly from the CD-ROM)
- Sound Blaster™ or other compatible sound card
- Internal/external speaker(s); headphones (recommended for classroom settings)

PRINTERS

Minimum Requirements:
- Laser printer, ink-jet printer, or 24-pin dot-matrix printer

INSTALLING THE PROGRAM

. . . ON MACINTOSH-COMPATIBLE COMPUTERS

1. Place the CD-ROM in the CD-ROM drive.
2. A window will appear with the program's "Read Me" file and the **HST Level 1, 2, or 3 Installer** icon. Double-click this icon and follow the procedures on the screen.
3. For information concerning installation of individual Explorations, please see the "Read Me" document on this screen.
4. After installation is complete, an **HST Level 1, 2, or 3** folder will appear on your hard drive. Open the folder and double-click the **HST** icon to launch the program.

. . . ON IBM-COMPATIBLE COMPUTERS (WITH WINDOWS)

1. Place the CD-ROM in the CD-ROM drive.
2. Locate the **Install** and double-click it.
3. Once the program is installed, a window will appear with the program's "Read Me" file, the **HST1, HST2, or HST3** icon, and the **Uninstall HST1, HST2, or HST3** icon.
4. Launch the program by clicking the **HST1, HST2, or HST3** icon found in the directory or folder of Program Manager.

Note: *For details on how to do a custom installation of individual Explorations to improve performance, please see the "Read Me" file located on the CD-ROM.*

LOGGING ON

Logging on to *Holt Science and Technology Interactive Explorations* is quick and simple. After launching the program, a log-on display will ask your students if they are working as guests, as individuals, or as a group.

... AS AN INDIVIDUAL

To log on as an individual, a student follows this procedure:

1. The student clicks the **Individual** button. A dialog box will appear asking the student to type in his or her first and last name and the teacher's name and class or period.
2. The student clicks **Enter** or presses **Return.**
3. When the main menu appears, the student chooses an Exploration.

... AS A GROUP

To log on as a group, students follow this procedure:

1. Students click the **Group** button. Students are then asked to type in their one-word group name, the teacher's name and the class or period, and the names of the members of their group.
2. Students click **Enter** or press **Return.**
3. When the main menu appears, the group chooses an Exploration.

... AS A GUEST (Nonassessed Use of the Program)

The log-on contains a guest feature that allows you or your students to do an Exploration without engaging the assessment function of the program. In other words, what is completed using the guest feature will not be graded. To log on as a guest, a student clicks the **Guest** button. When the main menu appears, the student chooses an Exploration.

An Important Note: *Please be sure to consult the "Read Me" file found on the CD-ROM. There you will find important, up-to-date information concerning general technical issues and changes that may not be present in this guide. Likewise, an additional "Read Me" file is contained in the Assessment Tools folder of the CD-ROM. This file identifies any updates for using the assessment tools.*

WELCOME TO THE MAIN MENU

The Main Menu allows you to select the following items:

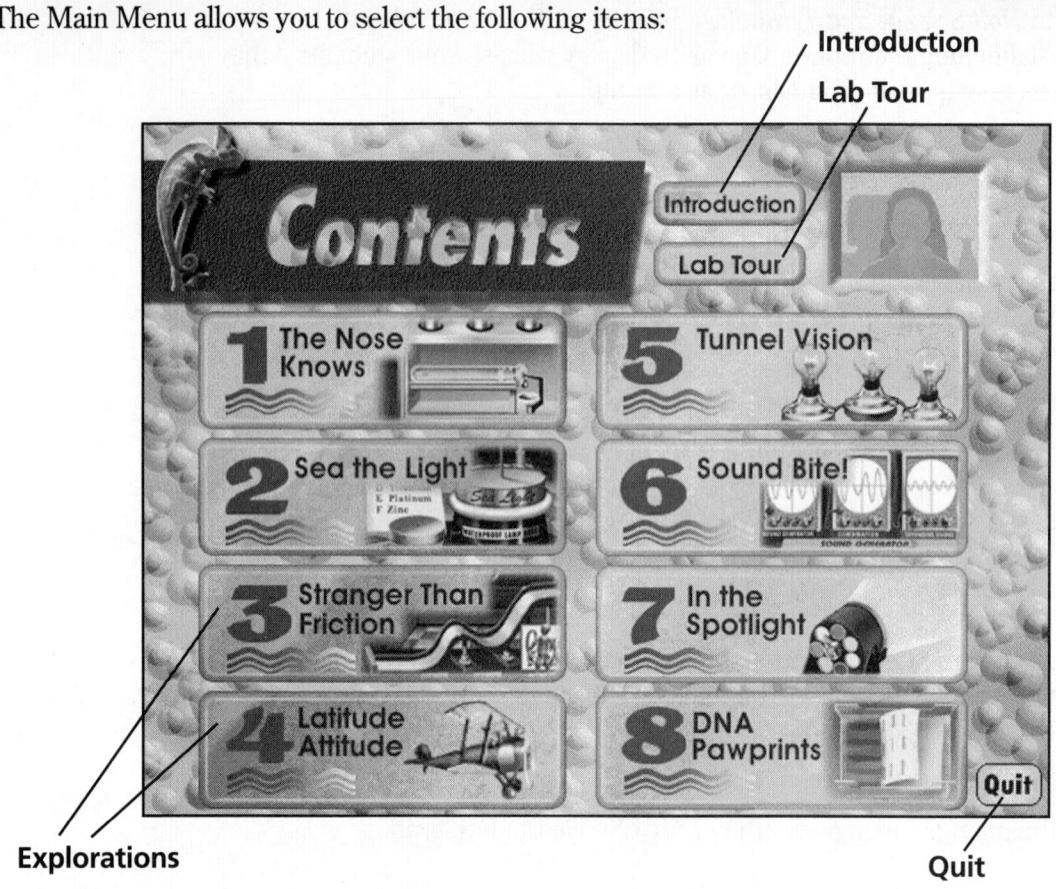

- **Introduction**

  Click the **Introduction** button to get a general overview of the Explorations.

- **Lab Tour**

  Click the **Lab Tour** button to get a quick yet comprehensive tour of Dr. Crystal Labcoat's laboratory. This is an excellent way for both you and your students to get acquainted with the standard equipment and features of the lab. Of course, depending on the Exploration and the problem to be solved, Dr. Labcoat has a wide variety of specialized equipment, which is described in the Exploration in which the equipment is used.

- **Interactive Explorations**

  Simply click any one of the eight **Exploration** buttons to start that particular Exploration. Should you want a quick overview of an Exploration, move the cursor over any Exploration button, and a brief description of the Exploration will automatically appear on the screen.

- **Quit**

  You can exit the program by clicking the **Quit** button on the main menu. Within an Exploration, you can also end your session by selecting **Quit** in the pull-down menu under **File**.

USING THE VIRTUAL LABORATORY

Dr. Labcoat's laboratory is a rich and functional scientific setting where students can practice their scientific problem-solving skills as well as their process skills as they try to solve a variety of science-related problems and mysteries. The following information will help you navigate and use the features of this unique laboratory.

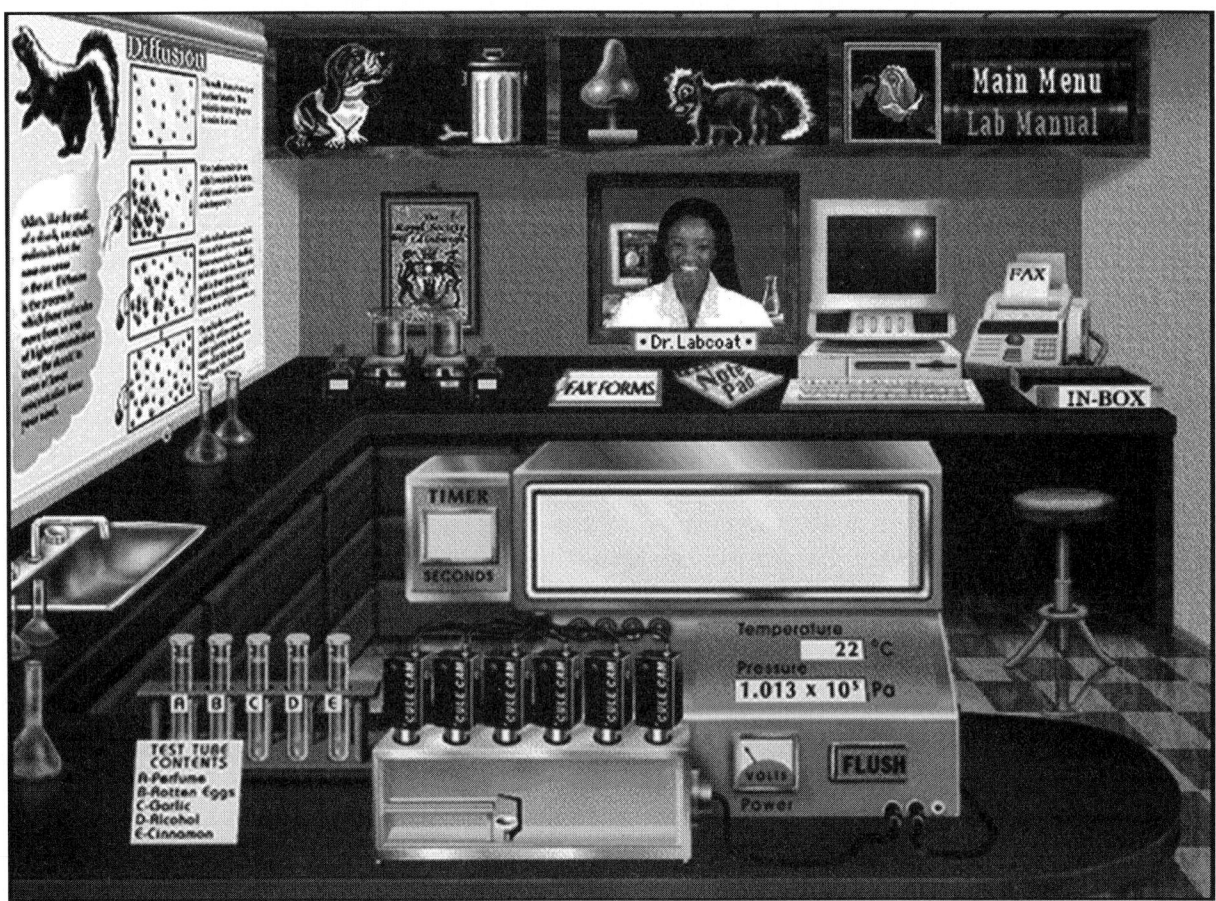

This virtual laboratory is from Disc 3, Exploration 1, The Nose Knows.

NAVIGATION

Navigation is accomplished by moving the mouse, which operates the cursor. You will notice that four types of cursors are used in the Explorations.

Arrow Cursor	↖	This point-and-click cursor is used to close pop-up windows as well as to set variables in experiments that require adjustments before a simulation can begin.
Pointing-Finger Cursor	☝	This point-and-click cursor indicates the areas of the lab that are active. Activate a selection by clicking it.
Hand Cursor	✋	This cursor indicates movable objects. To move objects, click and hold the mouse. You will be able to drag objects to lab equipment for testing as well as for classification and storage.
Bar Cursor	I	This point-and-click cursor appears in fields that require typing or word processing.

USER'S GUIDE   xvii

TOOL BAR

A tool bar featuring pull-down menus will display the following items:

File	Edit	Sound	Windows	Options
Quit	Copy Select All **Active when you are copying text from the Computer Database**	Level 0 Level 1 Level 2 Level 3 Level 4 Level 5 Level 6 Level 7	Notepad **Access to other pop-up variable panels available in certain Explorations**	English Lab Manual Audio Spanish Lab Manual Audio

LAB MANUAL

The Lab Manual provides a short explanation of the purpose and the operation of each piece of equipment in the lab. This is a handy reference for students who may need additional help.

To page through the Lab Manual, click the tabs at the lower right corner of the page. You can play audio instructions in English by clicking the megaphone-shaped icon. To hear instructions in Spanish, click **Options** on the tool bar and then select **Spanish Lab Manual Audio.** To hide the Lab Manual, click the **Close** box in the upper right corner.

FAX MACHINE

The Fax Machine is a primary means of communication both to and from the lab. Incoming fax messages identify problems to be solved. Outgoing fax messages are generated by students as they solve the problems and are asked to communicate their findings. Faxes often contain many pages. To page through faxes, click the tabs at the lower right corner of the page. To hide faxes, click the **Close** box in the upper right corner.

IN-BOX

The In-Box is where fax messages and other correspondence are kept for reference at any time during the Exploration. To page through faxes and other correspondence, click the tabs at the lower right corner of the page. To hide the correspondence, click the **Close** box in the upper right corner.

NOTEPAD

The Notepad is always available for jotting down notes and observations. Simply click the Notepad and start taking notes. Students can even paste articles from the Computer Database into the Notepad. Since the contents of the Notepad are not saved by the program, students should print their Notepads before leaving an Exploration if they are interested in keeping a record of their notes.

To paste text from the Computer Database into the Notepad, simply highlight the desired passages (or click **Select All**) and then select **Copy** from the pull-down **Edit** menu. Go to the Notepad and click the **Paste** icon. The text that you copied from the Computer Database will appear in the Notepad. **To print the contents** of the Notepad, simply click the **Print** icon.

More advanced students may choose to take notes or paste information from the Computer Database into other documents (such as those created by SimpleText) by running a separate word-processing application. Although this method is more complex, it allows students to save their notes electronically. For information concerning the simultaneous use of two applications, refer to the user's guide that accompanies your system's software.

COMPUTER DATABASE

Students can use this Computer Database to easily access information on a variety of subjects. The information is organized into articles that contain text as well as illustrations, photographs, and video. The Computer Database contains information that is vital to solving the problem or mystery.

To access articles that are relevant to an Exploration at hand, students simply scroll to the applicable topic and click on subtopics to view that information. Students can also access the complete database of articles by clicking the **Database Index** button. To return to the Exploration from which they accessed the database, students click **Table of Contents.**

It is important to note that students can copy text from the Computer Database into their notepads or other word-processing documents for printing. Images and video appearing in the Computer Database, however, cannot be pasted into the Notepad.

FAX FORMS

Fax forms located on the clipboard contain the forms necessary to create a new fax. Sending a fax is how students will communicate their solutions to the problems. Once a fax form has been completed, it is sent by clicking the **Send It** button. An appropriate response will be forthcoming from both the requester of the information and from Dr. Labcoat.

CALCULATOR

In Explorations that require mathematical calculations, a calculator is provided in the lab. Simply click the calculator to bring up the calculator's keypad.

ASSESSMENT TOOLS

A variety of assessment tools are available to make grading and record keeping as easy and as simple as possible. The assessment tools are located in the following folders, depending on which level CD-ROM you are using.

HST Level	Folder Name
Level 1	**adminHS1**
Level 2	**adminHS2**
Level 3	**adminHS3**

If you are using a Macintosh, you can access the Assessment Tools folder by entering the Preferences folder of your System folder. **If you are using a PC,** you will find the Assessment Tools folder in the Windows folder. You can also do a search for the Assessment Tools folder by entering the folder name as identified in the chart above.

When you open an Assessment Tools folder, you will find the following folders:

- Read Me
- Student Reports
- Answer Keys
- File Management

To view the contents of any folder, simply double-click the folder.

READ ME FOLDER

All updates to the assessment tools component of the program can be found in this printable file.

STUDENT REPORTS FOLDER

You can access student reports after each class session or at the end of the day by double-clicking the folder labeled "Student Reports." In the folder, you will see your students' work in the form of SimpleText files. These documents can be opened in SimpleText or in another word-processing application, if you choose.

Each student file or record has a prefix consisting of up to five letters. In the case of individual students, the prefix consists of the first four letters of the student's last name and the first initial of the student's first name; in the case of groups, the group's name (up to five letters) will appear. The prefix is followed by a period (.), and then a three-letter suffix (indicating the level of *Holt Science and Technology Interactive Explorations* and the Exploration number).

Consider the following examples:

File Name	Individual or Group Name	Level	Exploration
Smitj.L11	Smith, Jennifer	1	1
Smitp.L12	Smith, Peter	1	2
Aces.L27	Aces	2	7
Diamo.L38	Diamonds	3	8

All student records relate directly to the fax forms of the *Holt Science and Technology Interactive Explorations,* where students produce their work. Each student report contains the following sections:

Student/Class Information Section
This section contains:
- Student name or group name with names of group members
- Teacher's name
- Class or period
- The date the Exploration was conducted
- The duration of the student's work within the Exploration

Computer-Graded Section
This section contains the student's responses to the close-ended questions on the fax form. These questions, which are indicated with an asterisk (*), will have already been graded by the computer. Fifty (50) points are possible.

Teacher-Graded Section
This section consists of the answers to the open-ended questions on the fax form. This section must be read and evaluated by the teacher. Fifty (50) points are possible.

Scoring Section
This section consists of three scores: the computer-graded score, the teacher-graded score, and the composite score (final grade). Suggestions for grading student work can be found in the Answer Keys folder.

Teacher Comments Section
This section provides a space for making comments on your students' work. Simply type in your comments as you view the Student Report, or you can print out the Student Report and write in your comments manually.

ANSWER KEYS FOLDER
The Answer Keys folder contains eight answer keys, which correspond to the eight Explorations for a particular level. The files are named as indicated in the chart below. Please note that this chart only shows the file names for Exploration 1 in levels 1, 2, and 3. For Exploration 2, the file names would end in "2."

File Name	Level	Exploration
AnkeyL1.001	1	1
AnkeyL2.005	2	5
AnkeyL3.008	3	8

If desired, you can customize the names of these files simply by renaming them. Also, if you have installed two levels on the same machine, their respective contents will be found in another folder on your hard drive.

To use an answer key:

1. Double-click the answer key you wish to view.
2. Resize the answer key if you would like to view it on-screen along with a student report.
3. Grade your students' responses to the open-ended questions on the student report. Remember, the computer-graded questions are scored automatically.
4. If desired, print out hard copies of the answer key, the student's report, or both.

FILE MANAGEMENT FOLDER

The File Management folder contains an example of how you might set up class folders for storing your files. You can use this simple folder structure by changing the names of folders or adding more folders to suit your needs.

We suggest that you set up your records by class. At the end of each class, move student files into their corresponding class folder. You may decide to set up folders within each class folder by student or by Exploration number. Viewing files by date and time will be helpful in determining the class folder into which a student file can be inserted.

If you need help structuring your folders, please consult your computer manual for more in-depth instructions.

NETWORKING STUDENT REPORTS

The student reports in the Assessment Tools folder can be networked. This allows you to gather student reports to a central computer. To do this, you will need a dedicated network file server meeting the following requirements:

Macintosh-Compatible Computers

Minimum Requirements:
- 68040 CPU running at 33 MHz or higher, such as a Quadra or PowerPC
- 16 MB of RAM
- Recommended: 20 K per user per semester (or 40 K per user per year)

IBM-Compatible Computers

Minimum Requirements:
- Minimum 486DX running at 33 MHz or higher
- 16 MB of RAM
- Recommended: 20 K per user per semester (or 40 K per user per year)

You will also need Ethernet cards and cables, Novell Netware® 3.0 or higher, Windows NT™ 3.51 or higher, or AppleShare® 3.0 or higher network operating system.

OPTIMIZING PERFORMANCE

There are many things you can do to optimize the performance of *Holt Science and Technology Interactive Explorations*. Consider the following list of options:

1. Make sure that your monitor's resolution is set to 256 colors. Running the program on "Thousands of Colors" may slow the program down.

2. Do a custom installation of an Exploration rather than running the program completely off the CD-ROM. The program will perform better if an Exploration is installed onto your hard drive. Remove Explorations from the hard drive when you move on to a new Exploration.

3. If your hard drive is more than 80 percent full, performance may suffer. Remove old files and applications that are no longer pertinent or useful. Consider storing them on another hard drive or with alternative methods of data storage (such as Zip™ cartridges).

4. Use a utility to optimize your hard drive. Refer to the manual that came with your computer for more information about optimizing your hard drive.

5. If you are running other applications in the background—no matter how simple or complex the application—performance will suffer. Quit all applications other than *Holt Science and Technology Interactive Explorations*.

6. If you are on a network or on the Internet, your chances of experiencing freezes, crashes, and poor video or audio are much greater than if you are not.

7. Make sure you have the most current version of QuickTime 2.1 and QuickTime Powerplug extensions and the latest version of Sound Manager extension (v. 3.2). Using older versions of these extensions may result in reduced video quality.

8. If you have been running any memory-intensive applications prior to running *Holt Science and Technology Interactive Explorations*, restart your machine before running the program.

9. If the total memory requirement of your system software and this program approaches the limits of your machine's total RAM, performance will suffer. Consider upgrading your machine's capability by adding more RAM. Also consider using an extensions-management utility to turn off extensions not being used by the application or temporarily placing unused extensions in a folder labeled "Disabled Extensions."

10. Turn off your computer's Virtual Memory.

Suggestion: *If students are working in a setting that includes lower-end machines as well as newer models, you may want to rotate students or groups from one machine to another. Have your students do their research with the Computer Database section of the program on lower-end machines and do the main experiments using the better performing machines.*

TECHNICAL SUPPORT INFORMATION

At Holt, Rinehart and Winston we recognize the importance of providing you with the answers and help you need to use our quality instructional-technology products to their fullest potential.

Because systems, technology, and content are often inseparable, HRW has assembled a team of dedicated technical and teaching professionals and a suite of comprehensive support services to provide you with the support you deserve, 24 hours a day, 7 days a week.

Technical Support Line (800) 323-9239

The HRW Technical Support Line, which operates from 7 A.M. to 6 P.M. Central Standard Time, Monday through Friday, puts you in touch with trained Support Analysts who can assist you with technical and instructional questions on all of HRW's instructional technology products.

Technical Support on the World Wide Web http://www.hrwtechsupport.com

Contact the HRW Technical Support Center 7 days a week, 24 hours a day, at our site on the World Wide Web. Simply select the product you are interested in, and with a click of the mouse you can receive comprehensive solutions documents, answers to the most frequently asked questions, product specifications and technical requirements, and program updates from our FTP site. You can also contact our analysts at the Support Center using the following E-mail address: **tsc@hrwtechsupport.com**

Technical Support via Fax (800) 352-1680

Get the solutions you need with the HRW Technical Support Center's fax-on-demand service. Simply give us a call at our toll-free number to receive product-specific solutions within minutes. Our fax-on-demand service is available 7 days a week, 24 hours a day.

Exploration 1
Teacher's Notes

Something's Fishy

Key Concepts	A controlled experiment is an efficient way of determining how individual environmental variables affect living things. Acidic water conditions affect the health of African Cichlids.
Summary	African Cichlids are dying at the Fishorama store. The manager, Mr. McMullet, needs to know what changes he should make in his tanks to save the fish.
Mission	Find out why the African Cichlids are dying.
Solution	The pieces of ornamental driftwood that Mr. McMullet has been using in his aquariums have been leaching tannins into the water. Tannins are compounds that make the water more acidic. Such a change can be devastating to the health of the organisms that live in the water. The African Cichlids have been dying as a result of the water's increased acidity.
Background	Although difficult to care for, cichlids are interesting aquarium fish because of their bright colors. Unlike many other fishes, which abandon their offspring at birth, cichlids protect their young. Many aquarists enjoy watching the adult fish accompany the young cichlids around the tank.
	Scientists find the cichlids of African lakes particularly interesting because they speciate (develop into new species) so rapidly. In Lake Victoria, as many as 300 different species of cichlids have evolved since the lake formed 750,000 years ago. Such rapid evolution generally takes millions of years, and scientists are investigating environmental factors that could be causing this evolutionary explosion. Scientists are also concerned about how cichlid species are affected by human activities. Because many cichlid species are confined to lakes, they are especially vulnerable to environmental changes. The more we understand about the environmental changes that affect cichlids, the better our chances of ensuring their survival.

Exploration 1 Teacher's Notes, continued

Teaching Strategies

One purpose of this Exploration is for students to discover how to perform a controlled experiment effectively. Ideally, students will learn to do so by examining the CD-ROM articles provided and by analyzing their own successes and mistakes with the experiment. If necessary, review the CD-ROM articles about the scientific method with students who require additional assistance. Make sure students understand that observing a change in the experimental tank is only part of their goal. Isolating the specific variable responsible for that change is the most important part of their task. By comparing results from the experimental tank with results from the control tank, the effect of the variable can be determined. Changing the control tank in addition to the experimental tank would make it impossible to determine which variable is responsible for the results.

As an extension of this Exploration, you might wish to have students research the effects that changes in pH levels can have on species in lakes, streams, and rivers. Encourage students to find out more about the effects of acid precipitation on aquatic environments and to explore what is being done to protect wildlife in those areas.

Bibliography for Teachers

Ciresi, Rita. "One Fish, Two Fish, Red Fish, Blue Fish." *Penn State Agriculture,* Winter 1990, pp. 33–35.

Morris, Ronald. *Acid Toxicity and Aquatic Animals.* Cambridge, MA: Cambridge University Press, 1989.

Wilson, Edward O. *The Diversity of Life.* Cambridge, MA: Belknap Press of Harvard University Press, 1992.

Bibliography for Students

Encyclopedia of Aquatic Life. Keith Banister and Andrew Campbell, ed. New York, NY: Facts on File, 1985.

Kluger, Jeffrey. "Go Fish." *Discover,* 7 (18): March 1992, p. 18.

Other Media

Cichlid CD-ROM
Infobase Press
1844 S. Columbia Lane
Orem, UT 84058
801-221-1117

In addition to the above CD-ROM, students may find relevant information about cichlids on the Internet. Interested students can search for articles with keywords such as the following: *cichlids, aquariums, pets, hobbies,* and *tropical fish.*

Note: Remind any students who are interested in adding cichlids to their home aquariums that cichlids are difficult to care for. Encourage students to carefully research the special needs of cichlids before attempting to add them to their aquariums.

Name _____ Date _____ Class _____

Exploration 1
Worksheet

Something's Fishy

1. What kinds of problems is Mr. McMullet having with his African Cichlids?

   _____
   _____
   _____
   _____

2. What are five variables that might be affecting the African Cichlids?

   a. _____
   b. _____
   c. _____
   d. _____
   e. _____

3. What will you use for a control as you conduct your investigations?

   _____
   _____

4. Why is this control necessary?

   _____
   _____
   _____

5. Would it be better to test one variable at a time or several variables at once? Why?

   _____
   _____
   _____

6. Form a hypothesis for each of the experiments you conduct.

   Hypothesis 1: _____
   _____

EXPLORATION 1 • SOMETHING'S FISHY

Name _____ Date _____ Class _____

Exploration 1 Worksheet, continued

Hypothesis 2: _____

_____

Hypothesis 3: _____

_____

Hypothesis 4: _____

_____

Hypothesis 5: _____

_____

7. Record your observations as you investigate each hypothesis.

Hypothesis	Observations
1	
2	
3	
4	
5	

8. Were your experiments faulty in any way? If so, what steps did you take to correct your experiments?

_____

_____

_____

Record your conclusions in the fax to Mr. McMullet.

Name _____ Date _____ Class _____

Exploration 1
Fax Form

DISC 1

FAX

To: Mr. Ray McMullet (FAX 512-555-8633)

From:

Date:

Subject: African Cichlids

What is your recommendation? _____
_____

✂ ✂

For Internal Use Only

Please answer the following questions for my laboratory records. Scientists must always keep good records. *Dr. Crystal Labcoat*

During your experiments, which ONE of the following changes had a positive effect on the fish?

EXPERIMENTAL VARIABLES

☐ FEED FISH   ☐ INCREASE TEMPERATURE
☐ TURN LIGHT OFF   ☐ CHANGE FILTER
☐ REMOVE ORNAMENTAL DRIFTWOOD

Please explain why the African Cichlids responded to the above change.

_____
_____
_____
_____
_____

What effect did the change that you made have on the fish?

_____
_____
_____
_____

EXPLORATION 1 • SOMETHING'S FISHY

Something's Fishy

The following articles can also be found by accessing the computer graphic of the CD-ROM for Exploration 1:

- *Freshwater Aquariums*
- *Fish for Freshwater Aquariums*
- *Controlled Experiments*

Freshwater Aquariums

Keeping tropical fish is a way to learn about a wide variety of fish species and their habitats. Aquariums provide a small ecosystem that is not only rewarding to create but also enjoyable to observe.

Hardware

The hardware needed to keep freshwater tropical fish is relatively simple and inexpensive. Basically, all you need is a tank, a filter system, and a water heater. In most cases, new hardware for a 10-gallon tank will cost around $60 to $75.

Tank—Aquarium tanks range in size from small 5-gallon tanks to large tanks that contain 150 gallons or more. Most beginners start with a 10- to 20-gallon tank. The tank should be made specifically for tropical fish and should be equipped with a lid with a light.

Filter System—Healthy fish require water that is free of visible particles, dissolved carbon dioxide, and excess ammonia. This is achieved in an aquarium with a filter system. Box-type filters are designed to circulate the water through a layer of spun glass and activated charcoal to clean the water. Biological filters rely on colonies of bacteria to keep the water clean. The most common biological filter is the under-gravel filter. This filter is little more than a grate placed underneath the gravel at the bottom of the tank. In operation, water is pulled through the gravel into the grate and then up through tubes at each corner of the tank. Colonies of bacteria living in the gravel break down the contaminants in the water.

Many box-type filters and all under-gravel filters require an air pump to circulate the water through the filter system. The air pump also helps add oxygen to the water.

Water Heater—Because most tropical fish come from areas near the equator, they require warm water with little temperature variation. A water heater is used to ensure that the aquarium water stays at a warm, uniform temperature.

Healthy Water for Fish

In most cases, freshwater aquariums can be filled with tap water. But the water must be treated first to remove chlorine and chloramine. These chemicals are added to municipal water supplies to kill harmful microorganisms, but these chemicals can also kill tropical fish. Inexpensive chemicals that will effectively dechlorinate water are available at pet stores.

The water should also be kept at a stable pH level. The pH level indicates the number of hydrogen ions in the solution. The pH scale ranges from 0 to 14. Acids, such as vinegar and lemon juice, have a pH below 7. Bases, such as ammonia, have a pH above 7. Most freshwater fish can survive a pH range of 6.5 to 7.5. Most tap water falls within this range.

The pH of your aquarium water can change, however. The waste products from fish are high in ammonia and can quickly alter the pH of the water if the filter system is not functioning properly. Having too many fish in the tank or feeding them too much food can also cause excess ammonia and a high pH. In addition, some decorations can make tank water more acidic. Coral, shells, or bits of limestone can raise the pH. Ornamental driftwood may leach plant tannins into the water and lower the water's pH.

Test papers are compared with color charts for pH readings.

Preparing the Water for Fish

If a biological filter is used, the tank should be set up and filled with dechlorinated water 2–4 weeks before adding any more than one or two hardy fish. This will allow time for bacteria to colonize the gravel at the bottom of the tank. Two species of bacteria are necessary: one species that converts ammonia to nitrites, and another species that converts the nitrites to relatively harmless nitrates. After a few weeks, the water should have a high level of nitrates and little or no ammonia. Additional fish can then be added.

Water is tested for ammonia, nitrites, and nitrates since high levels of these substances can harm fish.

Fish for Freshwater Aquariums

General Health

Before purchasing any fish, inspect them carefully. Make sure they do not have unusual spots, raised and flaking scales, or other signs of disease or parasites. A 10-gallon tank can hold about 20 small- to medium-sized fish. To reduce stress on the fish, add only 3–5 fish to the tank at any one time. Give these fish 1–2 weeks to adjust to the tank. It normally takes about 2 months to completely stock a new aquarium with healthy fish.

Tetras

Tetras are very hardy fish. These fish come from Africa, Central America, and South America. Tetras prefer to live in schools of at least 6 fish. Several species require acidic water (pH about 6.5), so you should consider the pH of your tap water before you invest in these fish. If you want to keep tetras but your water is not acidic, you might try the black tetra or the flame tetra.

Cichlids

Native to Africa, Asia, Central America, and South America, the cichlids (SIK lids) are popular fish among aquarium owners. African Cichlids are extremely territorial and require places to hide, such as among plants, rocks, and wood. They also require water that is alkaline. Some cichlids live comfortably with other fish species, but other cichlids may eat other fish in the tank. Given the right conditions, cichlids can be some of the most colorful and most active fish in the tank. Some cichlids keep their bright colors only if they eat foods that contain certain orange, red, and yellow pigments, while others may lose their bright colors when exposed to conditions that are less than ideal.

Loaches

Loaches are excellent scavengers. They scour the bottom of the tank and eat food that has collected there. Native to Asia, they have long bodies and require little special care. Some popular species include skunk loaches, clown loaches, and blue loaches. Loaches prefer living with 2 or 3 of their own species.

Guppies

Guppies have beautiful fanlike tails that are speckled with a wide variety of colors and patterns. Most species of guppy are easy to raise. A few, however, require higher salt levels in water than most other fish can withstand. Guppies are unusual because they hatch from eggs kept inside the mother's body.

Angelfish

There are many species of marine angelfish but only one species of freshwater angelfish. This species is very hardy and attractive in an aquarium, but it can be aggressive with other fish.

Mollies

Two common varieties of mollies are the black molly and the sail-fin molly. Both are very hardy and beautiful fish to keep in an aquarium.

Plecostomus

The plecostomus is a scavenger fish. It eats algae on the sides of the aquarium and uneaten food that settles to the bottom of the aquarium. Take care when choosing a plecostomus for your aquarium; many varieties quickly outgrow their tanks. The clown and bristlenose plecostomus are good varieties for small aquariums.

Corydoras Catfish

Like the plecostomus, catfish are hardy and easy to raise scavenger fish. Corydoras catfish scour the bottom of the tank for food. Special sinking foods may be required to keep them well fed. Frozen worms or sinking pellets are a good food source for these catfish.

Controlled Experiments

Scientific Method

The scientific method is a systematic way of asking questions, performing experiments, gathering data, drawing conclusions, and communicating results. A scientist begins by asking an investigative question, such as "How do birds know when it's time to migrate?" or "Do heavier objects fall faster than light ones?" Then he or she collects information or data about the question to form a hypothesis. The hypothesis is a possible explanation for an event. A good hypothesis is a statement or explanation that can be tested. The scientist then designs an experiment to test the hypothesis. As the experiment takes place, observations are recorded. By analyzing these observations, the scientist can draw conclusions about the hypothesis and communicate the results.

What if you wanted to find out how much fertilizer is best for growing a potted plant? One way to do this scientifically is to set up a controlled experiment. A controlled experiment is one in which only one factor or variable is changed at a time.

In this case, the only variable that changes is the amount of fertilizer each plant receives. Everything else—including the type and size of plant, amount of water, amount and intensity of sunlight, and type of soil—must stay the same. The experimental plants would be exactly like the control plant, except they would be given different amounts of fertilizer. This way, you can easily test the effect of fertilizer on the growth of the plant.

Recording Results

Scientists must carefully record the results of their work. You may decide to record your results in your notepad using a form that looks something like this.

Title of the experiment:

Description of the problem or question:

Research about the problem or question:

Hypothesis:

Variables to be controlled:

Experimental variable:

Description of experiment performed:

Data and observations:

Do these observations support the hypothesis?

Additional questions or possible hypotheses:

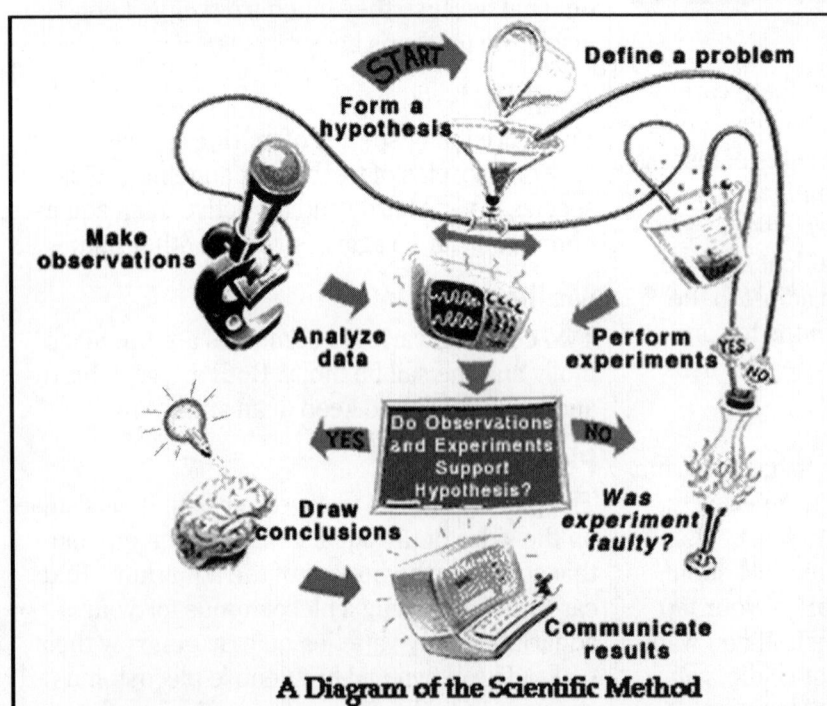

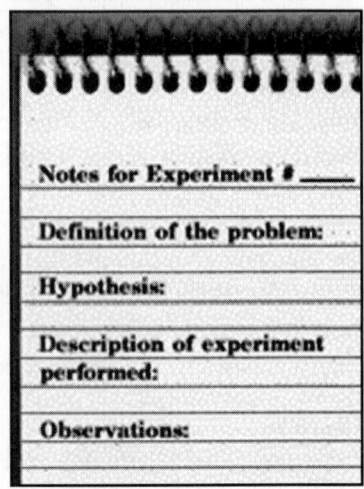

Scientists make careful notes of their observations and experimental results.

Exploration 2
Teacher's Notes

Shut Your Trap!

Key Concepts	Several variables may affect the health of living things. The Venus' flytrap gets nearly all the nutrients it needs from the insects it consumes; exposure to nutrient-rich soil will kill the plant.
Summary	Plant poachers have been taking Venus' flytraps from their native habitat in the bogs of North and South Carolina. The Bogs Are Beautiful Appreciation Society wants to prevent the Venus' flytrap from becoming extinct in the wild. The society needs to know the optimal growing conditions for the Venus' flytrap so that it can provide nursery owners with an easy and inexpensive method of growing the plant. Hopefully, the practice of poaching will then be reduced.
Mission	Determine the optimal growing conditions for the Venus' flytrap.
Solution	Several variables affect the growth of the Venus' flytrap. Nursery-cultivated flytraps thrive in conditions that most closely approximate the conditions of their natural habitat: 15 hours of light, 50 percent humidity, and no plant food. Adding plant food to the soil provides flytraps with an overabundance of nitrogen that kills the plants.
Background	The Venus' flytrap was once found in 21 counties surrounding Wilmington, North Carolina. Now, due to poaching and the destruction of the flytrap's natural habitat, the plant is found in only 11 of those counties. Even though flytraps can be grown in artificial environments, collecting the plants in the wild is cheaper, and poaching remains a problem. Because flytraps grow in sandy, boggy soil, they depend on periodic wildfires to clear patches in the bog for them to grow. However, the quick suppression of these fires by humans is contributing to the shortage of areas in which the flytraps can grow. Despite the work of government officials and the imposition of heavy fines for poaching, some scientists think that the flytraps may soon be found only in nature preserves and other protected areas.

Exploration 2 Teacher's Notes, continued

Teaching Strategies

This Exploration gives students the opportunity to conduct a controlled experiment in which several variables are combined to create optimal environmental conditions for the growth of a plant. Students must work efficiently because they have enough flies for only 10 experiments. Therefore, you may wish to encourage students to research the CD-ROM articles thoroughly before setting the variables in their control. Students who are having difficulty may require additional assistance. Make sure that they understand the purpose of a control. If necessary, review the diagram of the scientific method in the CD-ROM articles with students who are having difficulty.

As an extension of this Exploration, you may wish to have students discuss the consequences of allowing plant species to become extinct. Emphasize to students that finding an inexpensive method of growing the Venus' flytrap in an artificial environment could help to keep the plant from disappearing from the wild. Students may be interested to know that only 3 to 5 percent of the original habitat for carnivorous plants remains in the United States and that many species of carnivorous plants are already extinct. You might wish to have students contact an expert from a local college, university, or scientific institution to find out more about the preservation of disappearing plantlife.

Bibliography for Teachers

Albert, Victor A., Stephen E. Williams, and Mark William Chase. "Carnivorous Plants: Phylogeny and Structural Evolution," *Science,* 257 (5076): September 11, 1992, pp. 1491–1495.

Schnell, Donald E. *Carnivorous Plants of the United States and Canada.* Winston-Salem, NC: John F. Blair, Publisher, 1976.

Bibliography for Students

Doyle, Mycol. *Killer Plants: Venus' Flytrap, Strangler Fig, and Other Predatory Plants.* Los Angeles, CA: Lowell House, 1993.

Kite, L. Patricia. *Insect-Eating Plants.* Brookfield, CT: Millbrook Press, 1995.

Other Media

Death Trap
Videotape
Time-Life Video
777 Duke St.
Alexandria, VA 22314
703-838-7000
800-621-7026

In addition to the above videotape, students may find relevant information about the Venus' flytrap on the Internet. Interested students can search for articles with keywords such as *plants, carnivorous plants, insect-eating plants,* and *Venus' flytrap.*

Name _____ Date _____ Class _____

Shut Your Trap!

1. What are Ms. Lily N. Lotus and the Bogs Are Beautiful Appreciation Society concerned about?

   _____
   _____
   _____
   _____

2. What are three variables that may be affecting the growth of the Venus' flytraps?

   a. _____

   b. _____

   c. _____

3. What will you use for a control in your investigations?

   _____
   _____

4. There are 24 possible variable settings for the experimental terrarium. However, you have only enough flies to conduct 10 experiments. What steps can you take to make sure that you find a solution before you run out of flies?

   _____
   _____
   _____

5. Form a hypothesis for how each variable affects the growth of the Venus' flytraps.

   Hypothesis 1: _____

   _____

   Hypothesis 2: _____

   _____

   Hypothesis 3: _____

   _____

EXPLORATION 2 • SHUT YOUR TRAP! 11

Name _____ Date _____ Class _____

Exploration 2 Worksheet, continued

6. Record your observations in the table below as you investigate each hypothesis.

Plant food	Humidity	Hours of light	Observations
Yes	0%	5	
		10	
		15	
		20	
	25%	5	
		10	
		15	
		20	
	50%	5	
		10	
		15	
		20	
No	0%	5	
		10	
		15	
		20	
	25%	5	
		10	
		15	
		20	
	50%	5	
		10	
		15	
		20	

Name _____ Date _____ Class _____

Exploration 2 Worksheet, continued

7. Were your experiments faulty in any way? If so, what steps did you take to correct them?

_____
_____
_____
_____
_____
_____
_____
_____
_____

Record your conclusions in the fax to Lily N. Lotus.

Name _____ Date _____ Class _____

Exploration 2
Fax Form

FAX

To: Lily N. Lotus (FAX 910-555-5657)

From:

Date:

Subject: Optimal Growing Conditions for Venus' Flytraps

What is your recommendation? _____
_____
_____

For Internal Use Only

Please answer the following questions for my laboratory records. Scientists must always keep good records. Dr. Crystal Labcoat

During your experiments, which values proved to be optimal for the Venus' flytrap?

Optimal Values (select one per row)   **EXPERIMENTAL VARIABLES**

5	10	15	20	HOURS OF LIGHT PER DAY
0	25	50		PERCENT HUMIDITY
YES	NO			ADD PLANT FOOD?

What effect did the hours of light per day have on the plants? Why?
_____
_____
_____

What effect did the percent humidity have on the plants? Why?
_____
_____

14   HOLT SCIENCE AND TECHNOLOGY INTERACTIVE EXPLORATIONS TEACHER'S GUIDE

Name _____ Date _____ Class _____

Exploration 2 Fax Form, continued

What effect did the plant food have on the plants? Why?

_____

_____

_____

_____

_____

_____

_____

_____

_____

_____

Shut Your Trap!

The following articles can also be found by accessing the computer graphic of the CD-ROM for Exploration 2:

- *Venus' Flytrap*
- *Controlled Experiments*

Exploration 2
CD-ROM Articles

Venus' Flytrap

What a Peculiar Plant!

Close to the ground in a North Carolina bog, a predator lies in wait. Only 15 cm high, the Venus' flytrap lures its prey—ants, flies, and other small insects—into a deadly trap. The Venus' flytrap spreads its leaves open, and when an unsuspecting insect touches hairs on the leaf, the leaf snaps shut. The more the insect struggles, the tighter the trap becomes! Once the trap has closed, the plant releases digestive enzymes that slowly dissolve the prey.

A fully grown Venus' flytrap may have 6–15 traps that are about 2.5 cm long.

How does the trap work? Each leaf has two sides that are lined with sensitive trigger hairs. If an insect touches two of these hairs, or touches the same hair twice in a short period of time, the two sides snap shut. A change in water pressure causes the leaves to close together. When the trigger hairs are touched, water moves from cells on the inside walls of the leaves to cells on the outside walls of the leaves. The inside of the trap becomes limp and causes the leaf to close.

At Home in the Bogs

The natural habitat of the Venus' flytrap is the sandy, boggy soil that lies between the pine woodlands and the dense bogs of North and South Carolina. It has adapted itself to the unique environment of this region. This environment provides the Venus' flytrap with plenty of water, sunlight, and air.

If you want to raise a Venus' flytrap on your own, you must carefully regulate its environment. You can grow a Venus' flytrap in a sphagnum-moss mixture or in a large tray of water. The roots need a lot of room, so make sure your young plant has space to grow. Always use plenty of distilled water or unpolluted rainwater; tap water will quickly kill a Venus' flytrap. The plant thrives in high humidity, so you should try to maintain a consistent humidity level above 30 percent. In addition, the Venus' flytrap is adapted to a warm and sunny environment. Your plant will need strong light and a temperature that stays between 21°C and 38°C.

Why Eat Bugs?

The Venus' flytrap is a green plant. That means it makes its food through photosynthesis like other plants. So why does it trap and digest insects? The answer lies in the nutrient-poor soil. The wetlands soil lacks essential nutrients, especially nitrogen, that the plant obtains from the insects it digests. Like most other carnivorous plants, the Venus' flytrap cannot survive in nutrient-rich soils; its insect-eating adaptations allow it to thrive only in nutrient-poor soil.

Reproduction: Survival of the Species

In ideal growing conditions, a Venus' flytrap can bloom and produce seeds in about 3 months. First a stalk grows out of the center of the plant. It can reach about 15 cm in height. The stalk then sprouts pink and white flowers that allow the Venus' flytrap to reproduce. When the plant begins to flower, its traps shrivel up. At the base of the flower, green seedpods form which are eventually released into the soil.

As the coastal plain of the Carolinas develops, the existence of the Venus' flytrap becomes more threatened. Poachers, cashing in on this popular plant, are seriously diminishing native populations. While the Venus' flytrap once grew wild in 21 counties, now the species survives in just 11 counties. Laws in North Carolina inflict stiff penalties on poachers, but still the species is in danger.

Controlled Experiments

Scientific Method

The scientific method is a systematic way of asking questions, performing experiments, gathering data, drawing conclusions, and communicating results. A scientist begins by asking an investigative question, such as "How do birds know when it's time to migrate?" or "Do heavier objects fall faster than light ones?" Then he or she collects information or data about the question to form a hypothesis. The hypothesis is a possible explanation for an event. A good hypothesis is a statement or explanation that can be tested. The scientist then designs an experiment to test the hypothesis. As the experiment takes place, observations are recorded. By analyzing these observations, the scientist can draw conclusions about the hypothesis and communicate the results.

What if you wanted to find out how much fertilizer is best for growing a potted plant? One way to do this scientifically is to set up a controlled experiment. A controlled experiment is one in which only one factor or variable is changed at a time.

In this case, the only variable that changes is the amount of fertilizer each plant receives. Everything else—including the type and size of plant, amount of water, amount and intensity of sunlight, and type of soil—must stay the same. The experimental plants would be exactly like the control plant, except they would be given different amounts of fertilizer. This way, you can easily test the effect of fertilizer on the growth of the plant.

Recording Results

Scientists must carefully record the results of their work. You may decide to record your results in your notepad using a form that looks something like this.

Title of the experiment:
Description of the problem or question:
Research about the problem or question:
Hypothesis:
Variables to be controlled:
Experimental variable:
Description of experiment performed:
Data and observations:
Do these observations support the hypothesis?
Additional questions or possible hypotheses:

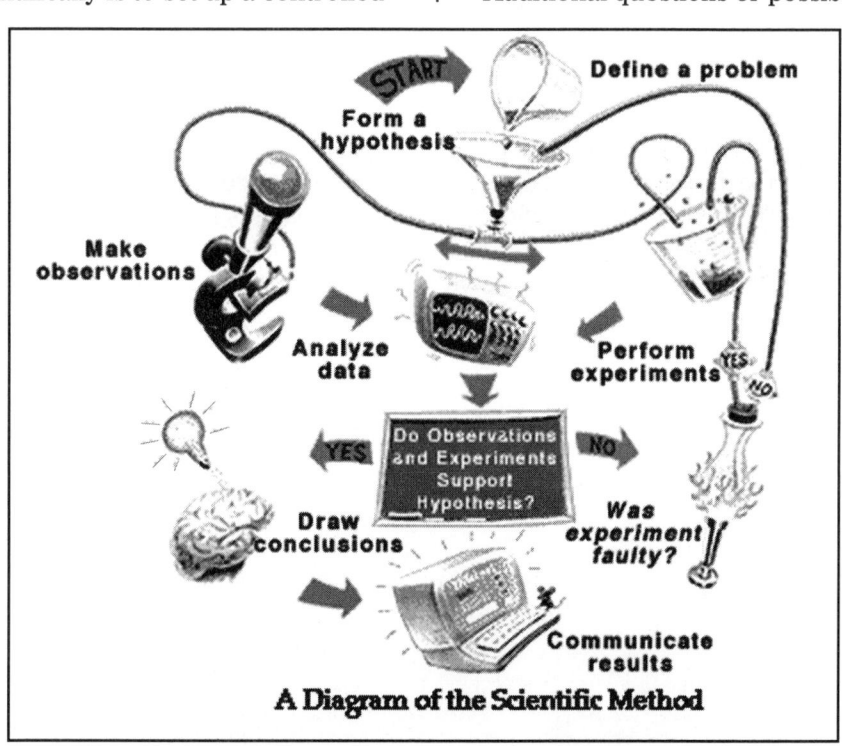

A Diagram of the Scientific Method

EXPLORATION 2 • SHUT YOUR TRAP!   17

Exploration 3
Teacher's Notes

Scope It Out!

Key Concepts	The spores of microorganisms can remain dormant for millions of years. Microorganisms can have helpful roles in the lives of other living creatures.
Summary	Dr. Viola Russ and her associates have been conducting experiments on amber. They have successfully extracted spores from a microorganism in a bee that was entombed in amber 25 million years ago. Dr. Russ now needs help to identify the microorganisms and their function and to discover whether they can be used to make antibiotics that will help fight modern diseases.
Mission	Identify the ancient microorganisms and determine their likely role in the life of the ancient bee.
Solution	The ancient microorganisms are rod-shaped bacteria of the kingdom Monera. They probably represent an ancient form of *Bacillus sphaericus,* a type of rod-shaped bacteria that aids bees in digestion and in fighting disease.
Background	Some students may ask why Dr. Russ is studying an ancient microorganism for use as an antibiotic. Explain that antibiotics are substances produced by microorganisms. Antibiotic substances are obtained from bacteria and fungi that live in the air, soil, and water. Doctors use antibiotics to fight various diseases that are caused by some harmful microorganisms. These diseases include tuberculosis, meningitis, pneumonia, and scarlet fever. Antibiotics are especially useful because they target specific cells without damaging others. Doctors can therefore use antibiotics to destroy harmful microorganisms in the body without affecting the body's cells. As a result, many diseases that were once fatal are now treatable with antibiotics. Scientists like Dr. Russ are interested in studying newly identified microorganisms because of the potential to develop new life-saving antibiotics.

Exploration 3 Teacher's Notes, continued

Teaching Strategies

Successful completion of this Exploration depends heavily on students researching the CD-ROM articles thoroughly. You may wish to encourage students to make use of the Notepad function as they conduct their research. In addition, student groups may benefit from a blackline master version of the articles. Those articles are available on pages 24–26 of this guide. Recommending that students reread the CD-ROM articles once the ancient microorganism has been identified may be especially helpful.

As an extension of this Exploration, you may wish to have students find out more about how to limit the harmful effects of microorganisms. For example, interested students could research the processes of pasteurization and sterilization and report their findings to the class in the form of an oral report. Students might also wish to contact a local restaurant to find out what steps are taken by the kitchen staff to prevent harmful microorganisms from spoiling food.

Bibliography for Teachers

Bodanis, David. *The Secret Garden: Dawn to Dusk in the Astonishing Hidden World of the Garden.* New York City, NY: Simon and Schuster, 1992.

Sagan, Dorion, and Lynn Margulis. *Garden of Microbial Delights: A Practical Guide to the Subvisible World.* Boston, MA: Harcourt Brace Jovanovich, Publishers, 1988.

Bibliography for Students

Dixon, Bernard. *Power Unseen: How Microbes Rule the World.* New York City, NY: W.H. Freeman & Company Limited, 1994.

Lovett, Sarah. *Extremely Weird Micro Monsters.* Santa Fe, NM: John Muir Publications, 1993.

Other Media

Organizing Protists and Fungi
Software (Apple II, MS-DOS, or Macintosh)
Queue, Inc.
338 Commerce Dr.
Fairfield, CT 06432
203-335-0906
800-232-2224

Simple Organisms: Bacteria (rev.)
Film and videotape
Coronet/MTI
P. O. Box 2649
Columbus, OH 43216
800-777-8100

Interested students may find relevant information about microorganisms and their roles on the Internet. Suggest that students search the Internet with keywords such as the following: *microorganisms, antibiotics, bacteria,* and *fungus.*

EXPLORATION 3 • SCOPE IT OUT! 19

Name _____ Date _____ Class _____

Exploration 3
Worksheet

Scope It Out!

1. What does Dr. Viola Russ need to know about the ancient microorganisms?

2. What does Dr. Russ intend to do with the results?

3. What will you use to conduct your investigation?

4. What do the ancient microorganisms look like under the microscope?

20 HOLT SCIENCE AND TECHNOLOGY INTERACTIVE EXPLORATIONS TEACHER'S GUIDE

Name _____ Date _____ Class _____

Exploration 3 Worksheet, continued

5. Use the table below to record your observations of each slide of microorganisms. Make sure that you write out the name of each microorganism in the left-hand column.

Protista	Observations
1.	
2.	
3.	
4.	
Monera	**Observations**
5.	
6.	
7.	
Fungi	**Observations**
8.	
9.	
10.	

EXPLORATION 3 • SCOPE IT OUT! 21

Name _____ Date _____ Class _____

Exploration 3 Worksheet, continued

6. Which modern microorganisms look the most like the ancient microorganisms?

   _____
   _____
   _____

7. How might you classify the ancient microorganisms and find out more about their likely role in the life of the ancient bee?

   _____
   _____
   _____
   _____
   _____
   _____
   _____
   _____
   _____
   _____

Record your conclusions in the fax to Dr. Russ.

Name _____ Date _____ Class _____

Exploration 3
Fax Form

FAX

To: Dr. Viola Russ (FAX 805-555-2266)

From:

Date:

Subject: Ancient microorganism classification and probable function

What are the ancient microorganism's classification and function? _____
_____
_____
_____

For Internal Use Only

Please answer the following questions for my laboratory records. Scientists must always keep good records. Dr. Crystal Labcoat

Which of the following may be used to classify the ancient microorganisms? Place an X in the left-hand column beside the correct answer(s).

KINGDOM	PROTISTA	MONERA	FUNGI
	Euglena	Round-shaped bacteria	Mildew
	Paramecium	Rod-shaped bacteria	Mold
	Amoeba	Spiral-shaped bacteria	Yeast
	Algae		

What role did this microorganism most likely play in the life of the ancient bee?

_____
_____
_____

EXPLORATION 3 • SCOPE IT OUT! 23

> **Scope It Out!**
>
> The following articles can also be found by accessing the computer graphic of the CD-ROM for Exploration 3:
>
> • *The World of Microorganisms*

Exploration 3
CD-ROM Articles

The World of Microorganisms

What Is a Microorganism?

Microorganisms are so small that they can be seen only with a microscope. Microorganisms can be found in almost every environment on Earth. They can be found in air, soil, water, and other living things. Many microorganisms form the basis of food chains. Some cause diseases, such as malaria, AIDS, and strep throat. Other microorganisms help larger organisms to survive by aiding in digestion and other life processes.

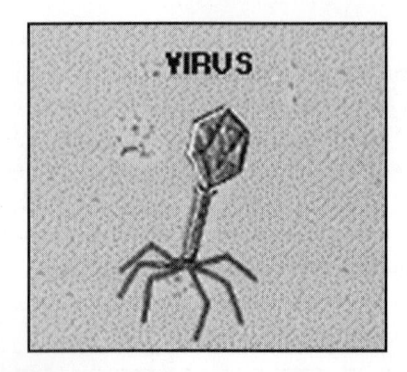

Viruses

Among the smallest of microorganisms, viruses can be seen only with a powerful electron microscope. Because viruses cannot reproduce without infecting living cells, cannot make proteins, cannot use energy, and are not made of cells, biologists do not consider viruses to be living organisms. However, viruses can harm living organisms and in fact are responsible for a number of diseases. Viruses are made up of strands of genetic material inside a protein coat. The information from a cell's genetic material is responsible for all of the cell's life processes.

When a virus infects a living cell, it replaces the cell's genetic information with its own, which causes the cell to stop functioning properly.

Viruses cause many common diseases, such as the cold, influenza (the flu), and chickenpox. Generally, the human body can defend itself against these infections. Some viral diseases can be prevented by immunizations, which are usually injections of weak or dead viruses that "trick" the body into behaving as if it has already had the disease. That way, the body may be able to quickly fight off a dangerous, invading virus in the future. Other viral diseases are more deadly. HIV is the virus that causes AIDS. HIV attacks the body's immune system, making it difficult for the body to defend itself against infection. Researchers continue to search for a vaccination against HIV and a cure for AIDS.

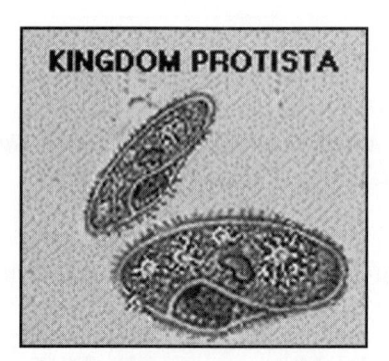

Kingdom Protista

Most organisms that belong to the kingdom Protista are single-celled. These organisms have some of the characteristics of plants and some of the characteristics of animals. They all live in moist environments, and they display a wide variety of shapes and structures. Consider the protists described below.

Amoeba—Amoebas are found in fresh water, salt water, and soil. They move by stretching out pseudopodia of intercellular liquid. Pseudopodia bulge out from the rest of the amoeba and can engulf food particles for the amoeba to digest. Amoebas reproduce by dividing in two. Most live freely in water or soil, while some, such as the amoeba that causes dysentery, infect other organisms.

Dinoflagellate—This microorganism is found in the ocean. It moves by using two flagella that cause it to spin through the water like a top. Some produce poisons that can harm other marine organisms, but most dinoflagellates are an abundant source of food for other species. A few of these kinds of organisms give off a glow much like fireflies!

Euglena—The single-celled Euglena can make its own food like a plant or absorb nutrients from its environment. The Euglena lives in fresh water and moves by wiggling its tail, which is called a flagellum. The Euglena has an eyespot that helps it orient its movements toward light.

Green algae—Most green algae live in fresh water and are single-celled; some, however, are multicellular. Green algae contain the same pigments that are found in most plants. For this reason, many scientists believe that green algae are the ancestors of the plant kingdom.

Paramecium—The Paramecium lives in fresh water and uses thousands of fine hairs, called cilia, to move and feed. The cilia beat in waves that cause the Paramecium to spin as it moves through the water. The cilia also sweep food into a narrow groove where the food is engulfed by a small, liquid-filled bubble called a vacuole. As the vacuole moves through the Paramecium, the food inside it is digested.

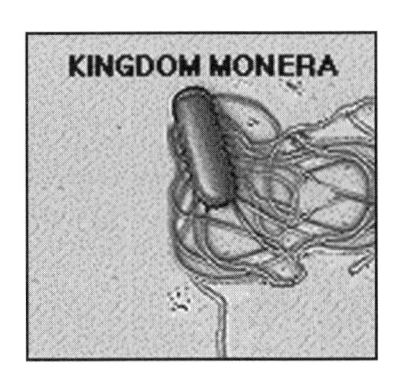

Kingdom Monera

The most common monerans are bacteria, which are single-celled microorganisms that do not have a nucleus but do have a cell wall. They are the oldest, simplest, and most abundant form of life on Earth. Bacteria can live in a variety of conditions, from hot springs to icy mountaintops. Some bacteria form spores in which most life functions stop until environmental conditions are able to support the bacteria again. Bacteria are classified according to their shape. Cocci are round, bacilli are rod-shaped, and spirilla are shaped like spirals. Bacteria reproduce quickly by dividing into two cells. Under proper conditions, one bacterium can divide every 20 minutes. Some forms of bacteria cause disease either by destroying living cells or producing dangerous toxins. Strep throat is a common bacterial disease caused by streptococcus bacteria.

Other bacteria perform useful functions, such as breaking down dead matter and other wastes. Bacteria are even used in producing foods such as cheese and yogurt.

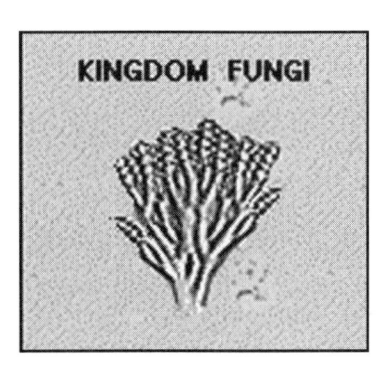

Kingdom Fungi

Fungi are microorganisms that break down once-living material and absorb food through their cell walls. The bodies of fungi are made of a network of threadlike structures called hyphae. Reproductive spores are attached to the ends of the hyphae. Once released, these spores can survive for long periods of time before they find the proper environmental conditions and begin to divide. Most fungi need a moist environment to survive, and fungi frequently appear after a rainfall or in damp areas. Common kinds of fungi include mold, mildew, yeast, puffballs, and mushrooms. Mold appears as a fuzzy substance on the surface of once-living material, such as fruit or other foods. Mildew appears as a powdery or downy covering and often grows on leaves. Yeasts are a kind of fungi that form sacs of spores. Bakers use a kind of yeast to make bread rise. Puffballs and mushrooms form club-like structures in which the cap contains the spores.

Helpful or Harmful?

Are microorganisms helpful or harmful? The answer depends on the microorganism in question. Some microorganisms cause deadly diseases. The bubonic plague, typhoid, and cholera are caused by bacteria. AIDS, polio, measles, and mumps are caused by viruses. One form of dysentery is caused by a kind of amoeba.

Other microorganisms are helpful. *Penicillium* is a kind of fungi that fights bacterial disease. Yeast, another type of fungi, is used in baking and brewing. A certain bacteria, *Escherichia coli,* lives in the digestive tracts of humans and aids in the breakdown and digestion of food. Another bacteria, *Bacillus sphaericus,* plays a similar role in the digestive tracts of bees. Several forms of bacteria are used to make products such as cheese and yogurt.

Escherichia coli

Most microorganisms cannot be classified as helpful or harmful. These organisms are simply a part of the environment, carrying out roles that we have yet to learn much about.

Microorganisms Trapped in Time

In 1991, Dr. Raul Cano announced that he had revived microorganisms that were between 25 and 40 million years old. These microorganisms were trapped in the gut of a bee that had been preserved in amber. The amber formed when resin flowed from a tree and hardened. The bee died quickly in the amber, but the microorganisms inside its body did not. Instead, some of the microorganisms became dormant.

Dr. Cano says that he revived the dormant microorganisms by exposing them to the proper nutrient-rich environment. Critics of Dr. Cano's work do not believe that such ancient life can be revived. They say that modern microorganisms must have contaminated Dr. Cano's results. Dr. Cano contends that he followed strict guidelines to prevent any contamination. In addition to the microorganisms found in the ancient bee, Dr. Cano claims to have revived over 2,000 other kinds of microorganisms that were trapped in a variety of insects.

Exploration 4
Teacher's Notes

What's the Matter?

Key Concepts	Each element has its own unique melting point and its own unique boiling point. Experimental procedures should be planned in advance to avoid unproductive investigations.
Summary	Several researchers have been investigating the physical properties of molten lava, but, unfortunately, the tip of their temperature-measuring device, the lava analyzer, has melted. They need to know what type of metal will make the most practical replacement for the tip of the lava analyzer.
Mission	Determine which metal will make the best tip for the lava analyzer.
Solution	The high melting point of titanium, its availability, and its relatively affordable price make it the best metal to use for the tip of the lava analyzer. Tungsten and platinum are reasonable choices because both can withstand the heat of the molten lava. However, their scarcity and costliness make these metals less practical than titanium.
Background	Students may be interested to know that scientists are actually studying the Kilauea volcano in Hawaii. The Hawaiian Volcano Observatory (HVO) is located on the summit of Kilauea, one of the most active volcanoes in the world. Frequent eruptions at Kilauea and nearby Mauna Loa make the HVO an ideal location for studying volcanoes. As a result, many of the world's most famous vulcanologists have studied at this observatory. In addition, many important new techniques and instruments for monitoring volcanoes have originated at the HVO.
	In addition to the HVO, there are two other major sites in the United States for studying volcanoes, the Cascades Volcano Observatory and the Alaska Volcano Observatory. At all three of these sites, scientists monitor geophysical changes involving seismicity, ground movements, gas chemistry, hydrologic conditions, and the activity between and during eruptions. In addition, the scientists carefully analyze data to warn of possible eruptions and of the specific hazards related to those eruptions. When eruptions do occur, these scientists are fully prepared to study the eruptive behavior, identify the activities that led to the eruption, and define the processes by which its deposits are left behind.

Exploration 4 Teacher's Notes, continued

Teaching Strategies

One of the purposes of this Exploration is to allow students to make decisions about how to conduct an investigation in the most efficient manner possible. One activity, the determination of the mass and volume of each metal, is a distracter. With careful forethought, students may be able to predict ahead of time that such an activity would have no relevance to the task at hand. Encourage students to establish in advance a procedure for solving Dr. Stokes's problem. Be sure that students include a careful study of the CD-ROM articles in their procedure.

As an extension, you may wish to encourage students to do further research on one of the metals described in the Exploration. Students could focus their research on how the metal is used in its solid, liquid, and gaseous states in important everyday products. Suggest that students report their findings to the class in the form of a poster or other visual display.

Bibliography for Teachers

Barber, Jaqueline. *Solids, Liquids, and Gases: A School Assembly Program Presenter's Guide.* GEMS Series, edited by Lincoln Bergman and Kay Fairwell. Berkeley, CA: Lawrence Hall of Science, 1986.

Davies, Paul, and John Gribbin. *The Matter Myth.* New York City, NY: Simon & Schuster/Touchstone, 1992.

Bibliography for Students

Berger, Melvin. *Solids, Liquids and Gases: From Superconductors to the Ozone Layer.* New York City, NY: G.P. Putnam's Sons, 1989.

Cooper, Christopher. *Matter.* Eyewitness Science series. New York City, NY: Dorling Kindersley, Inc., 1992.

Darling, David. *From Glasses to Gases: The Science of Matter.* New York City, NY: Dillon Press, 1992.

Other Media

Particles in Motion: States of Matter
Filmstrip
National Geographic Society
Educational Services
P. O. Box 98019
Washington, DC 20090-8019
800-368-2728

Interested students may also find relevant information about states of matter on the Internet using keywords such as the following: *physical changes; matter;* and *solids, liquids, and gases.* Students may also be interested in Internet resources about *lava, volcanoes, vulcanologists,* and *volcano observatories.*

Name _____ Date _____ Class _____

Exploration 4
Worksheet

What's the Matter?

1. What problem does Dr. Stokes need you to help him solve?

   _____

   _____

   _____

2. Dr. Labcoat has gathered a set of metal samples for you to analyze. List the name of each metal and record its melting point and boiling point in the following data chart:

Metal	Name of metal	Melting point (°C)	Boiling point (°C)
Cu			
Sn			
Pt			
Ti			
W			
Al			

3. How might the information in the table be useful to you in solving Dr. Stokes's problem?

   _____

   _____

   _____

   _____

4. What additional information could help you solve the problem?

   _____

   _____

   _____

   _____

EXPLORATION 4 • WHAT'S THE MATTER?

Name _____ Date _____ Class _____

Exploration 4 Worksheet, continued

5. How might you find this information?

   _____
   _____
   _____
   _____

6. Record helpful data here, continuing on the back of the page if necessary.

   _____
   _____
   _____
   _____
   _____
   _____
   _____
   _____
   _____
   _____
   _____

7. When you have finished, evaluate the procedure that you used to complete this Exploration. What would change about your procedure? Did you perform any activities that were not useful to you? If so, which ones?

   _____
   _____
   _____
   _____

8. How could you improve your procedure?

   _____
   _____
   _____

Record your conclusions in the fax to Dr. Stokes.

Name _____ Date _____ Class _____

Exploration 4
Fax Form

FAX

To: John Stokes, Ph.D. (FAX 080-555-9822)

From:

Date:

Subject: Metal recommendation

What is your recommendation? _____
_____
_____
_____

✂·✂

For Internal Use Only

Please answer the following questions for my laboratory records. Scientists must always keep good records. *Dr. Crystal Labcoat*

Please indicate your metal selection here: _____

How do the particles of this metal behave during the following phases:

solid? _____
_____
_____

liquid? _____
_____
_____

gas? _____
_____
_____

EXPLORATION 4 • WHAT'S THE MATTER?

What's the Matter?

The following articles can also be found by accessing the computer graphic of the CD-ROM for Exploration 4:

- *Matter*
- *Metals*
- *Volcanoes*

Exploration 4
CD-ROM Articles

Three states of matter: solid, gas, and liquid

Matter

The Basics of Matter

Matter is defined as anything that takes up space and has mass. Matter is made up of small particles called atoms and molecules. Different atoms and molecules have different physical and chemical properties.

Odor, color, taste, hardness, mass, density, melting point, and boiling point are just a few examples of *physical properties*. All of these properties can be observed and even measured without changing the composition of the matter. For example, you can observe the melting point of an ice cube without the ice cube changing into a different kind of matter. This is not the case with chemical properties.

Chemical properties can be observed only when one kind of matter changes into another kind of matter. For example, when you add baking soda to vinegar, a different kind of matter is formed—carbon dioxide gas. A chemical property of baking soda, then, is the formation of a gas when vinegar is added.

States of Matter

Most matter you will encounter exists in one of three states: solid, liquid, or gas.

Solids are rigid, cannot be noticeably compressed, and usually have distinct boundaries. The atoms and molecules in solids are tightly arranged, like people sitting in rows in a theater.

Liquids flow, cannot be noticeably compressed, and usually have a boundary with air. In liquids, atoms and molecules move past each other but remain within a boundary. A liquid can be compared to people moving around in a theater's lobby.

Gases flow, can be compressed, and have no boundary with air. Atoms and molecules in gases spread apart, much like people leaving the theater and going in different directions.

Changing States of Matter

If you leave a candle sitting in the hot sun, it will change from a rigid stick to an oozing liquid. This transformation is called a *change of state* or a *phase change*. The candle changes state but does not change its composition. The oozing liquid is still wax. This means that the basic particles that make up the wax do not change but the form of the wax does. For another example, consider water. A water molecule remains a water molecule whether it is ice (solid), water (liquid), or vapor (gas). The water has simply undergone changes of state. And since these changes of state do not change the composition of the water, they are considered physical changes.

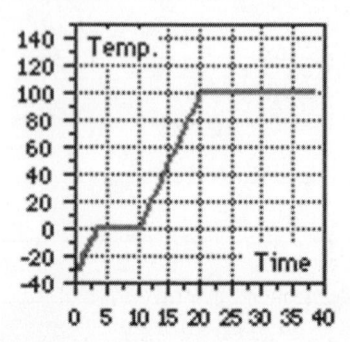

Graph showing the roles of temperature and time in changing states of matter

The state of a particular type of matter depends primarily on its temperature. When a solid is heated, its temperature rises until it reaches its melting point. The melting point is the temperature at which a solid changes to a liquid. The melting-point temperature remains constant until all of the solid has changed to a liquid. As the liquid is heated, the temperature begins to rise again until it reaches its boiling point. The boiling point is the temperature at which a liquid changes to gas. The temperature remains constant at the boiling-point until all of the liquid has changed to gas.

The Nature of Matter

The kinetic molecular theory of matter states that matter is composed of very tiny particles that are in constant motion. This theory helps explain some of the physical properties of solid, liquid, and gaseous matter.

The particle model of matter assumes four basic principles about matter.

1. All matter is composed of tiny particles called atoms.

2. Each element is made up of the same kind of atoms, and the atoms of one element are different from the atoms of all other elements.

3. Atoms cannot be divided, created, or destroyed.

4. Atoms of elements combine in certain ratios to form compounds.

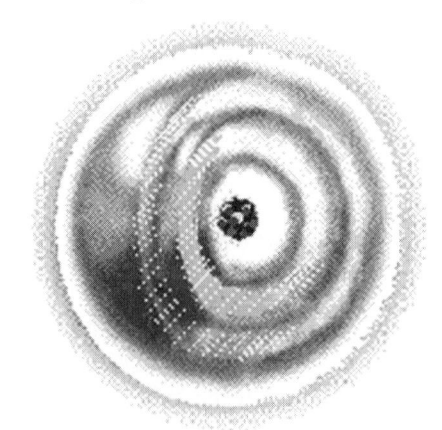

Model of an atom

Although modern research has shown that not all matter is composed of atoms and that atoms are composed of even smaller particles, this model is still a useful way to understand the nature of matter.

Periodic Table of Elements

The periodic table of elements is a chart that shows all of the known elements arranged by similar physical and chemical properties. Elements with similar properties are grouped together in vertical columns. Major classes of elements include the alkali metals, lanthanides, actinides, halogens, and noble gases. In addition to classifying elements by properties, the periodic table also helps scientists make predictions about the properties of new elements by comparing them to the properties of known elements already in the chart.

Each square on the chart describes an element by listing the element's atomic number, chemical symbol, element name, and atomic mass. The atomic number identifies the number of protons in an atom's nucleus. Atomic mass is the average mass of the element. The chemical symbol is an international abbreviation for the element. Sometimes the symbol is the first letter of the element's name, such as *C* for carbon. However, the symbol may also come from a Greek or Latin word for the element. The symbol for lead is Pb, which comes from *plumbum*, the Latin word for "lead." Some elements are named for a place or individual; for example, Einsteinium was named in honor of the physicist Albert Einstein.

Metals

Aluminum
Atomic Number: 13
Atomic Mass: 26.9815
Description: silvery white

Clays containing aluminum were used for making pottery in the Middle East over 7000 years ago. Other aluminum compounds were also used in ancient Egypt and Babylonia.

Aluminum is an extremely useful metal, primarily because it is lightweight. Despite this, aluminum mixtures can be made strong enough to replace many heavier metals. Aluminum is also long lasting and less expensive than many other metals. It conducts heat and electricity well and does not rust easily. Aluminum is an excellent reflector of heat, so it is useful in building insulation and roofing materials. Most aluminum produced today is made from bauxite ore, which makes up about 8 percent of the Earth's crust. Large deposits are found throughout the world.

It takes less energy to recycle used aluminum than to produce new aluminum. Recycling efforts provide over half the aluminum supply used in making new aluminum products, mostly aluminum cans. Aluminum is used in many other objects such as airplanes, automobile parts, household utensils, foil, and building supplies.

Copper

Atomic Number: 29
Atomic Mass: 63.546
Description: reddish

Copper is extracted from veins of copper ore in the Earth. Once refined, copper is quite malleable, which means it can be easily shaped or formed. People have used copper since prehistoric times to make tools, utensils, jewelry, and weapons. In modern times, copper is used in piping, cookware, wire, and coins. There are more than 160 known minerals that contain copper. One of these minerals is yellow chalcopyrite, otherwise known as fool's gold.

Copper is extremely useful because it readily conducts heat and electricity. Silver is the only metal that conducts electricity better than copper at room temperature. Copper also resists rust and is relatively inexpensive. It can be mixed with other metals to form strong alloys such as brass and bronze.

Gold

Atomic Number: 79
Atomic Mass: 196.967
Description: bright yellow, soft

Gold has been a symbol of wealth for thousands of years. It is scarce, but can be found worldwide. Scientists believe that geologic activity moved gold from the Earth's inner depths through cracks and fissures to the Earth's surface. This explains why gold is found in igneous rocks and sea water throughout the world. Deposits of gold are small and rarely pure. Gold is normally found mixed with other metals, such as silver, lead, or copper. Although gold is found in all sea water, its extraction from sea water is expensive.

Gold does not tarnish, rust, or corrode like most other metals. Gold is also the most malleable metal, and it can be hammered into sheets that are so thin they are transparent. Gold is often mixed with other metals, such as silver, copper, and nickel, to form alloys. The amount of gold in each of these alloys is measured in karats. Gold is most often used for decorative purposes, such as jewelry, but is also used in sophisticated electronics, such as those used on spacecraft and satellites.

Iridium

Atomic Number: 77
Atomic Mass: 192.22
Description: hard, brittle, silvery

Iridium is from the same family as platinum and is one of the hardest metals known. Ores and deposits containing pure iridium are extremely scarce, and separating it from other compounds is a very complicated and costly process. Iridium was named after the Greek goddess of the rainbow, Iris, because the salts of this metal are brightly colored. Generally, iridium is mixed with other metals to add hardness to the resulting alloy. Iridium is used to make jewelry, surgical instruments, pen points, and electrical contacts.

Iron

Atomic Number: 26
Atomic Mass: 55.847
Description: silvery white, lustrous, magnetizable

Iron composes about 5 percent of the Earth's crust, making it one of the most abundant elements. Since prehistoric times, iron has been extracted from its ores and shaped for a wide variety of uses. It is estimated that around 1200 B.C., people discovered how to melt iron and forge it into tools and weapons. This discovery began a period in history called the Iron Age.

Iron is still widely used today because it is plentiful and relatively inexpensive. Common items such as frying pans and pipes are made of iron. Iron can also be mixed with other metals to make very strong alloys, such as steel.

Another widely used iron alloy is cast iron, which is valuable because of its hardness, low cost, and ability to absorb shock. Cast iron is a compound that contains from 2 to 4 percent carbon and 1 to 3 percent silicon. Because of its high carbon content, cast iron can be shaped only by melting it and pouring the liquid metal into a mold. This casting process is used to make items such as automobile engines, fire hydrants, and construction materials.

Lead

Atomic Number: 82
Atomic Mass: 207.29
Description: silvery, soft

Lead is one of the heaviest metals known, but it is so soft that it can be scratched with a fingernail. It is easily molded into a wide variety of objects, such as pipes, sculptures, fishing weights, and bullets. Because lead is plentiful, it is an inexpensive metal to obtain.

Some lead compounds are poisonous, limiting its use in many products. Until the 1970s, lead was used as a white pigment in house paint. Many children became sick from eating the peeling paint chips that contained lead. Other substances, including titanium dioxide and zinc, are now used in paint instead of lead.

Platinum

Atomic Number: 78
Atomic Mass: 195.08
Description: silver-white

Platinum is found worldwide and is usually mixed with other metals. However, it is not an abundant metal. It makes up only about a millionth of 1 percent of the Earth's crust. Since deposits of platinum are so small, the price of this metal is similar to the high price of gold.

Platinum is easily shaped, does not tarnish, conducts electricity, and resists high temperatures and chemicals. These properties make it a useful metal for missile cones, jet-engine fuel nozzles, and dental fillings. In addition, platinum is often used for making jewelry. When combined with gold, it makes an alloy known as "white gold."

Silver

Atomic Number: 47
Atomic Mass: 107.868
Description: brilliant white

Silver has long been popular because of its beauty and usefulness. It can be found in veins of pure silver or, more commonly, with other elements. Although silver is considered a rare metal, it costs less than platinum or iridium.

For thousands of years, silver has been used as money. A soft, malleable metal, it is usually combined with other metals to make tableware, coins, and jewelry. High-grade silver jewelry is usually made of sterling silver, which is an alloy of silver and copper.

One of silver's most valuable properties is that it conducts heat and electricity better than any other known metal. For this reason, it is often used to make electrical contacts. Other industrial uses for silver include photographic film, batteries, and coatings for the backs of mirrors. Silver is also used to plate, or coat, less expensive metals.

Tin

Atomic Number: 50
Atomic Mass: 118.69
Description: silvery white

Many people associate tin with tin cans. But a tin can is actually a steel can coated with tin. Because tin is nontoxic, malleable, and resists corrosion, it makes an excellent protective coating for other metals. Tin is also used to make alloys such as bronze (tin and copper) and pewter (tin, copper, and antimony).

Tin is usually extracted from its main ore, cassiterite. Because tin occurs less frequently in nature than copper and aluminum, it is more expensive than these metals.

Titanium

Atomic Number: 22
Atomic Mass: 47.88
Description: strong, silvery gray

Titanium is the ninth most abundant element on Earth. Rock samples from the moon show that this metal is present there, too. Titanium is as strong as steel and 45 percent lighter. It resists corrosion and high temperatures, which makes it especially useful for airplanes and spacecraft. Most white paint now contains highly reflective titanium dioxide rather than more toxic lead compounds. Surgical-implant devices are also made of titanium and its alloys because they are compatible with human tissue.

Tungsten

Atomic Number: 74
Atomic Mass: 183.84
Description: hard, brittle, steel gray to white

Tungsten is the twenty-third most abundant element on Earth, which makes it less common than iron, titanium, or zinc. It has the highest melting point of any metal, making it ideal for high-temperature uses such as light-bulb filaments, fireproof cloth, and heating elements for electric furnaces. In its pure state, tungsten is very hard and brittle. Because of these qualities, it is often used in alloys to add hardness and strength to softer metals. However, tungsten is always found in an impure state in nature and is costly and difficult to process and refine.

Tungsten filament

Zinc

Atomic Number: 30
Atomic Mass: 65.39
Description: bluish white

Zinc ranks fourth, after steel, aluminum, and copper, among the metals most commonly produced. It replaces lead in paint because it is nontoxic, reflects light, and resists mold and fungus. Zinc is also mixed in small quantities with copper and other metals to make bronze and brass. Zinc is fairly inexpensive and is widely distributed throughout the world.

Zinc can be used as a protective coating that keeps steel from rusting. In a process called *hot-dip galvanizing*, the steel is heated, dipped into a vat of molten zinc, and then passed through a cooling tower, where the zinc coating hardens.

Common items made of galvanized steel include heating ducts, storage tanks, light poles, fencing, and highway guardrails.

Zirconium

Atomic Number: 40
Atomic Mass: 91.22
Description: grayish white

Zirconium is a rare, soft metal that comes from the mineral zircon, which often looks very much like diamonds. Zirconium is found only in very small quantities on Earth. Traces of it have also been identified in the sun, other stars, and some meteorites.

Because zirconium has a very high melting point and resists corrosion and the absorption of atomic particles, it is widely used in the field of nuclear energy. It is also used in rocket engines, high temperature furnaces, and medical hardware such as pins and screws used to repair broken bones.

Nuclear energy symbol

Volcanoes

Eruptions on Earth

A volcano is an opening in the Earth's surface through which gases and hot molten rock erupt. Molten rock, called magma, is pushed up through the Earth by explosive gases deep within the Earth's mantle. Sometimes when the magma reaches the Earth's surface, it is called lava.

Vulcanologists are scientists who study volcanoes. Vulcanologists estimate that there are 40,000 volcanoes on Earth, three-fourths of which are located under water. Volcanoes do not occur only on Earth, however. Scientists have discovered volcanoes on Mars and Venus, as well as on Io, and Europa, two of Jupiter's moons.

Exploration 4 CD-ROM articles, continued

Types of Volcanoes

Scientists classify volcanoes based on how the eruptions form mountains. In some eruptions, lava simply pours from an opening in the crust. Hardened lava gradually builds up, forming a mountain, or cone, with gently sloping sides. This type of volcano is called a *shield volcano*. Mona Kea, in Hawaii, is a shield volcano. Despite its gentle slope, it rises 10,203 m from the ocean floor to its summit, forming the largest mountain on Earth. Measured from its base to the ocean floor, it is almost 1400 m taller than Mount Everest.

Not all eruptions are as gentle as those that form shield volcanoes. When magma is thick, pressure from trapped gases within the magma may build up, causing an explosive eruption to take place. Such an explosion can spew lava and tephra—volcanic rock, cinders, and ash—several kilometers into the air. This eruption creates a *cinder-cone volcano,* which has steep sides and a narrow base. Once the pressure is released, this type of volcano usually becomes inactive. One example of a cinder-cone volcano is Paricutín, in Mexico. This volcano grew to 100 m in its first five days of eruption in 1943.

Some volcanoes, called *composite volcanoes,* are formed from alternating layers of lava and tephra. Many well-known volcanoes, such as Vesuvius and Mount St. Helens, are composite volcanoes. They usually have steep slopes and broad bases. Because of their layered construction, composite volcanoes are sometimes referred to as *stratovolcanoes.*

The eruption of a composite volcano

Lava

Lava varies in color from dark red to light yellow, depending on its chemical composition and temperature. The temperature of lava ranges from 600°C to 1200°C. Lava keeps heat so well that it may take up to a month for a 1 m thick flow to cool. As it cools, lava hardens into many different forms. Smooth, folded sheets of lava are called pahoehoe. Jagged, rough sheets are called aa. Lava also cools into domes, steep hills, and tunnels.

The Ring of Fire

Volcanoes are often located where the large rigid plates of Earth's crust meet. Many of these plates are located along the ocean floor. So many volcanoes encircle the Pacific Ocean plate that this region is called the Ring of Fire.

One of the volcanoes in this region is Kilauea (kil uh WAY uh), on the island of Hawaii. Kilauea lies far from a plate boundary, but at a place where a huge column of magma rises from within the Earth. Kilauea, which means "rising smoke cloud," has a deep, round crater at its center. When the volcano erupts, the crater fills with a lake of boiling lava.

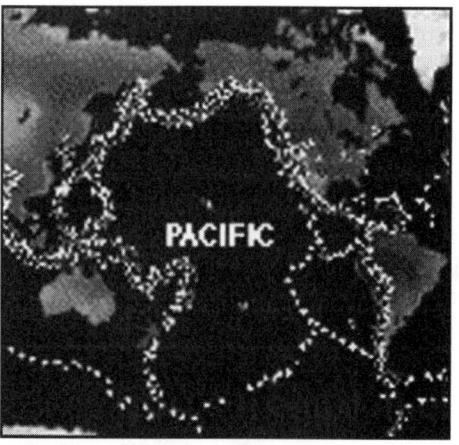

The Ring of Fire is a line of volcanic activity that circles the Pacific Ocean.

EXPLORATION 4 • WHAT'S THE MATTER?

Exploration 5
Teacher's Notes

Element of Surprise

Key Concepts	Elements in the same chemical group, or family, of the periodic table of elements have similar chemical properties. Predictions about the reactivity of an element may be based on the reactivity of another element in the same chemical group.
Summary	Mr. Fred Stamp of Pack & Mail, Inc. has to deliver some potentially dangerous chemicals to a remote research station on the continent of Antarctica. Because Mr. Stamp and his crew are going to be traveling via dog sled over and near water, he needs help determining the reactivity of each element to water. He has requested that each of the 12 elements that he will deliver be given a rating of extremely reactive, reactive, or not reactive to water. He will then use these ratings to construct the best possible transport containers for the elements.
Mission	Help to ensure that some potentially explosive chemical samples are safely delivered to the South Pole.
Solution	By testing the samples in the laboratory and by using their knowledge that elements in the same chemical group have similar chemical properties, students can determine the reactivity of the 12 samples. Krypton, like neon, radon, and xenon, is not reactive; magnesium, like calcium, barium, and strontium, is reactive; and sodium, like potassium, cesium, and rubidium, is extremely reactive.
Background	Shipping elements to a research team in Antarctica is not such a far-fetched idea. Antarctica is the continent about which we know the least. This vast region of unknowns lures scientists from all over the world to make new discoveries and to study everything from astrophysics to microbial ecology. In fact, the Antarctic Environmental Protection Act of 1996 designated Antarctica as a "natural reserve devoted to peace and science."
	Located almost completely within the Antarctic circle, Antarctica contains 90 percent of the world's ice. Because the ice sheets reflect most of the sun's heat back into the Earth's atmosphere, the South Pole region has an annual mean temperature of $-49°C$. Despite the extreme cold, scientists gather regularly at a number of antarctic research stations. At the South Pole Station, for example, astronomers from around the world study galaxy and star formation while astrophysicists keep an eye on the ozone layer. The McMurdo Dry Valleys are another hot spot for research on Antarctica. This unusual expanse of ice-free land, with its mountain ranges, meltwater streams, and arid terrain, draws botanists, geochemists, biologists, ecologists, and other scientists.

Exploration 5 Teacher's Notes, continued

Teaching Strategies

One likely outcome of this Exploration is that students will have a greater understanding of the fact that elements in the same chemical group of the periodic table share similar chemical properties. Students are more likely to stay focused on this project if they are reminded that in a real-life situation, an incorrect prediction about the reactivity of the elements could lead to serious injury or even death. As always, the students will find a good deal of helpful information in the CD-ROM articles, so be sure that students have read them carefully before reporting their findings to Mr. Stamp. If students have difficulty with this Exploration, encourage them to focus on what the tested chemicals in each reactive group have in common. Then direct their attention to the periodic table, and ask them what is similar about the placement in the table of the nonreactive, reactive, and highly reactive groups of elements.

As an extension, suggest that students research the types of containers that are appropriate for storing reactive, nonreactive, and highly reactive chemicals. Encourage the students to present their results in the form of a visual display. You may also wish to extend the Exploration with further analysis of other families in the periodic table. Divide the class into small groups, and provide each group with a list of several chemicals from the same family of the periodic table. Also include a description of the chemical properties of one of the chemicals. Then challenge the groups to make predictions about the chemical properties of the other chemicals in the list, based on the one description.

Bibliography for Teachers

Tocci, Salvatore, and Claudia Viehland. *Chemistry: Visualizing Matter.* Austin, TX: Holt, Rinehart and Winston, 1996.

Bibliography for Students

Cobb, Vicki. *Chemically Active: Experiments You Can Do at Home.* Philadelphia, PA: Lippincott, 1990.

Loesching, Louis V. *Simple Chemistry Experiments with Everyday Materials.* New York City, NY: Sterling Publishing Co., Inc., 1991.

Other Media

Chemistry: The Periodic Table and Periodicity
Film, videotape, and videodisc
Coronet/MTI Film & Video
P. O. Box 2649
Columbus, OH 43216
800-777-8100

Interested students may also find relevant information about elements on the Internet. Suggest that students use keywords such as the following to conduct their search: *chemistry, chemicals, chemical changes, elements,* and *reactivity.* Students may also wish to access interesting facts about *Antarctica* by exploring the Internet.

Name _____ Date _____ Class _____

Exploration 5 Worksheet

Element of Surprise

1. Mr. Stamp needs your help. Describe your assignment.

   _____
   _____
   _____
   _____

2. What materials are available in Dr. Labcoat's lab to help you complete your assignment?

   _____
   _____
   _____
   _____

3. Describe what you will do to test each element's reactivity to water.

   _____
   _____
   _____
   _____

4. Record your findings about each sample in the spaces that follow.

   a. barium: _____
   _____
   _____
   _____

   b. calcium: _____
   _____
   _____
   _____

40 HOLT SCIENCE AND TECHNOLOGY INTERACTIVE EXPLORATIONS TEACHER'S GUIDE

Name _____ Date _____ Class _____

Exploration 5 Worksheet, continued

   c. cesium: _____

   d. neon: _____

   e. potassium: _____

   f. radon: _____

   g. rubidium: _____

   h. strontium: _____

   i. xenon: _____

EXPLORATION 5 • ELEMENT OF SURPRISE

Name _____ Date _____ Class _____

Exploration 5 Worksheet, continued

5. What additional information do you need to complete your assignment (to determine the reactivity of krypton, magnesium, and sodium to water)?

   _____

   _____

   _____

   _____

6. Now that you know the reactivity of each of the 12 elements, how do you think Mr. Stamp should pack the chemicals when preparing to deliver them to Antarctica?

   _____

   _____

   _____

   _____

   _____

   _____

   _____

   _____

   _____

   _____

   _____

   _____

Record your conclusions in the fax to Mr. Stamp.

Name _____ Date _____ Class _____

Exploration 5
Fax Form

FAX

To: Mr. Fred Stamp (FAX 011-619-555-7669)

From:

Date:

Subject: Chemical Properties of Elements

Select the appropriate classification for each of the following chemicals:

CHEMICAL REACTIVITY WITH WATER

CHEMICAL	EXTREMELY REACTIVE	REACTIVE	NOT REACTIVE
BARIUM			
CALCIUM			
CESIUM			
NEON			
POTASSIUM			
RADON			
RUBIDIUM			
STRONTIUM			
XENON			

Please utilize the above information to predict the chemical reactivity of the following chemicals:

CHEMICAL	EXTREMELY REACTIVE	REACTIVE	NOT REACTIVE
KRYPTON			
MAGNESIUM			
SODIUM			

How did the periodic table help you to answer Mr. Stamp's questions?

_____

_____

_____

_____

DISC 1

EXPLORATION 5 • ELEMENT OF SURPRISE 43

Exploration 5
CD-ROM Articles

Element of Surprise

The following articles can also be found by accessing the computer graphic of the CD-ROM for Exploration 5:

- *The Periodic Table of the Elements*
- *Properties of Matter*
- *Antarctica*

The Periodic Table of the Elements

How It Began

In the 1800s, scientists were struggling with a problem. Many elements had been discovered and analyzed, but scientists could not agree on the best way to organize the elements. Dmitri Mendeleev, a Russian chemist, proposed a solution that would later become our modern periodic table of the elements.

In 1869, Mendeleev arranged the elements according to their atomic masses. The atomic mass of an element equals the sum of the masses of the protons and neutrons in one atom of the element. He organized the elements in rows, with each row containing elements with similar properties. When Mendeleev created the periodic table, he predicted the future discovery of certain elements and their properties based on gaps that remained in the chart. Over time, scientists were able to fill in the chart. As they did, they found that most of Mendeleev's predictions were correct.

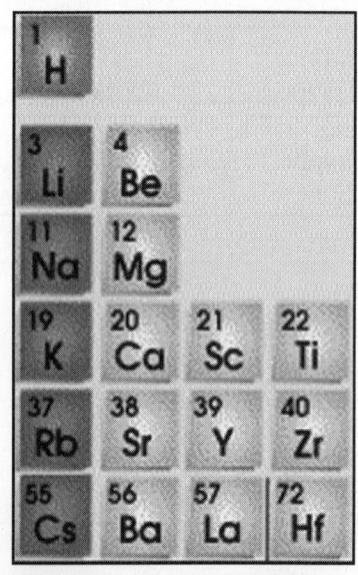

In the modern periodic table, atoms are arranged by atomic number—the number of protons in the nucleus of one atom of the element. Instead of organizing elements with similar properties in horizontal rows, the modern table organizes elements in columns. Scientists continue to modify the chart. And although all of the naturally occurring elements have been discovered, analyzed, and named, scientists can artificially create atoms with higher atomic numbers. These can then be analyzed, named, and placed on the chart.

Organizing the Elements—Trends in the Table

The modern periodic table summarizes a great deal of information about the elements. Each square on the table lists the name and symbol for the element, the atomic number, and the average atomic mass. In addition, the placement of each element on the table indicates some of the element's properties.

Elements are arranged by their increasing atomic numbers in the periodic table of the elements.

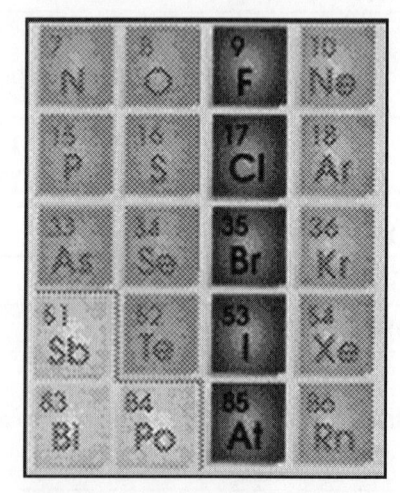

Elements in the same families have similar properties.

The atomic number establishes the order of the elements, but it is the number of electrons that determines the common properties of **groups** of elements. Groups are arranged in vertical columns on the table.

44 HOLT SCIENCE AND TECHNOLOGY INTERACTIVE EXPLORATIONS TEACHER'S GUIDE

An element's group is determined by the number of electrons in the unfilled shell, or energy level, of the atom. The first shell can contain up to 2 electrons. The second and subsequent shells can contain up to 8 electrons. For example, hydrogen has an atomic number of 1. It has 1 electron in its unfilled shell. Lithium has an atomic number of 3. It has 2 electrons in its first shell and 1 electron in its unfilled shell. Because both hydrogen and lithium have a single electron in the unfilled shell, they belong to the same group. Elements in the same group share similar reactive properties. Elements that have their electron shells filled belong to a group called **noble gases**. These gases are sometimes classified as inert, or not reactive.

A Host of Elements

The periodic table lists over 100 different known elements. The descriptions here will give you an idea of some of the many differences between elements.

Barium

Atomic number: 56
Physical properties: Soft, silvery white metal

Barium compounds are often used when a doctor administers an X ray. The patient swallows a barium compound before having the X ray. The compound absorbs the X rays and reveals a clear picture of the digestive tract.

Calcium

Atomic number: 20
Physical properties: Moderately hard, silvery metal

Calcium is found in bones and shells of living organisms.

Calcium makes up about 3 percent of the Earth's crust. It is a primary component in bones, shells, and many rocks, including limestone and gypsum.

Cesium

Atomic number: 55
Physical properties: Soft, silvery white metal; liquid at room temperature

Cesium reacts quickly when exposed to light, so it is used in photocells. Scientists also use cesium to measure radiation from cosmic rays and nuclear particles.

Krypton

Atomic number: 36
Physical properties: Whitish and gaseous

Krypton is often used with other gases, such as neon, in fluorescent lamps and neon signs. It produces a bright, orange-red glow when an electric current is passed through it.

Magnesium

Atomic number: 12
Physical properties: Moderately hard, silvery metal

Because magnesium burns with a dazzling white light, it is used in fireworks, flares, and flashbulbs. Magnesium is also a part of the chemical chlorophyll, which allows plants to make food. It is an important component in certain medicines, fertilizers, and cements.

Neon

Atomic number: 10
Physical properties: Colorless and gaseous

Neon is an extremely rare element found in the Earth's atmosphere. It glows with a bright orange-red light when an electric current passes through it.

Potassium

Atomic number: 19
Physical properties: Soft, silvery white metal

Potassium is usually combined with other elements. It is a major component of the Earth's crust and plays an important role in the nervous system of many animals. Extremely explosive, potassium is used in matches and fireworks. It also occurs in combination with other elements in soaps, fertilizers, and dyes.

Exploration 5 CD-ROM articles, continued

Radon
Atomic number: 86
Physical properties: Colorless, radioactive, and gaseous

Radon is formed from the disintegration of the element radium. Because it produces radiation, it is very dangerous to life-forms. Radon has been discovered in homes built over soil and rock that have high concentrations of radium.

Rubidium
Atomic number: 37
Physical properties: Soft, silvery white

Rubidium ignites in air and reacts violently in water. Like cesium, its high reactivity makes it useful in photocells.

Sodium
Atomic number: 11
Physical properties: Soft, silver-white metal; malleable

The sixth most abundant element in the Earth's crust, sodium is found in both soil and water. It regulates water and nerve function in living cells. Sodium and chlorine form table salt, a common chemical. Sodium is also an important ingredient in baking powder, lye, soap, fertilizers, and many other products.

Strontium
Atomic number: 38
Physical properties: Soft, silvery white metal

Strontium ignites in air and produces a brilliant red flame. It is often used in fireworks and flares. Strontium-90 is a form of strontium that is radioactive and dangerous to living things.

Xenon
Atomic number: 54
Physical properties: Colorless, odorless, and gaseous

Xenon is found in tiny quantities in the atmosphere. It is used in certain lamps and light bulbs because it can produce a bright light.

Properties of Matter

Physical and Chemical Properties of Matter

A physical property is a property of matter that can be observed without changing the composition of the matter. Mass, hardness, boiling point, melting point, color, and texture are all physical properties. Physical properties are often easy to observe. Consider the physical properties of pure water in a glass. In its liquid state, water is clear, odorless, and has a measurable mass. You do not have to change the composition of the water to make these observations. If you wanted to determine the water's freezing or boiling point, you still would not change the water's composition.

This is not the case with a chemical property. A chemical property is any property of matter that describes how one kind of matter interacts with other kinds of matter. When a piece of iron is left in the rain, for example, it reacts with the water to form iron oxide, or rust. When iron rusts, a chemical change occurs. The very composition of the iron changes. Because this chemical change occurs, a chemical property of iron is that it reacts with water to form iron oxide, or rust.

Each element has a unique set of physical and chemical properties. However, elements in the same group tend to share certain chemical properties. This occurs because many chemical properties are determined by the reactivity of the element. The reactivity of an element is largely based upon the number of electrons in the element's unfilled electron shell.

Antarctica

Antarctica is the Earth's most southern continent and largest icecap.

A History of the Continent

After studying fossils and geological formations, some scientists have speculated that Antarctica once had a very different climate and position on Earth. Over 200 million years ago, Antarctica may have been a part of Gondwanaland, a large, warm continent that joined all seven existing continents.

The continent of Antarctica, an area that measures approximately 13 million square kilometers, is almost entirely covered by a thick sheet of ice. The ice sheet was created by the buildup of snow over millions of years, and it contains over 70 percent of the world's freshwater supply. As the ice piles up, it turns into glaciers and ice rivers that flow from the continent to the sea. Often, large glaciers break off and float out to sea until they melt in warmer waters. A vast current called the Antarctic Circumpolar Current moves the cold waters into the rest of the world's oceans.

Exploration

Antarctica was the last continent to be discovered and explored. Many people wanted to reach the South Pole, which lies on the interior of the frozen land. The first expeditions that landed on the continent's coast were recorded in 1895. By 1901, British explorer Robert F. Scott began the first expedition to reach the geographic South Pole. He failed in his attempts and was beaten to the location by Norwegian Roald Amundsen in 1911. These early explorers moved across the continent using dog sleds, sails, and other primitive resources. Many of these adventurers died in the frozen climate.

By the late 1920s, airplanes made conquering the difficult terrain easier. In November 1929, American Richard Byrd flew over the geographic South Pole. He continued exploring Antarctica by air and on land. By the late 1950s, several nations had joined together to establish major scientific research centers at the geographic South Pole and around the continent.

The South Poles

Antarctica actually contains three points known as south poles. The geographic South Pole lies at a latitude of 90° south of the Earth's equator. The magnetic south pole is the location that compasses point toward. It is located to the north and east of the geographic South Pole. The geomagnetic south pole is the location of the Southern Hemisphere's auroras, which are brightly colored lights that appear in the sky. It lies between the geographic and magnetic poles.

EXPLORATION 5 • ELEMENT OF SURPRISE

Exploration 6
Teacher's Notes

The Generation Gap

Key Concepts	Alternative sources of electricity, such as wind and solar power, can be less expensive than coal, oil, and other common sources of electricity. A number of factors can affect the viability of wind-generated electricity.
Summary	Ms. Wendy Powers of EcoCabin, Inc. would like to know if it would be cost-effective to use the Electroprop wind turbine to provide electricity for a small log home in the San Francisco area. The turbines cost $14,560 and should not need repairs for 20 years. Ms. Powers would like to know how long it will take for the turbine to pay for itself.
Mission	Test a wind turbine to determine if it will lower the overall energy costs for a small log home.
Solution	If installed in the San Francisco area, the Electroprop wind turbine would save the homeowner an average of $1456 a year and would pay for itself in 10 years. Since the turbine should not need repairs for 20 years, the turbine should be considered highly cost-effective.
Background	Fossil fuel supplies are limited, and most people agree that we are using these resources much faster than they can be replaced by nature. For this reason, fossil fuels are considered to be nonrenewable resources. In addition to being limited, fossil fuels have other drawbacks. Oil spills, strip mining for coal, and exhaust from power plants and motor vehicles can all have negative effects on our environment. As a result, many nations today are focusing on developing new energy sources, making the most of the ones we already have, and using every form of energy as efficiently as possible. Many energy-wise plans focus on renewable resources, resources that are continually produced. Some renewable energy resources, such as wind and sunlight, are so abundant that they are considered to be inexhaustible.
	Tremendous energy is contained in the wind. This energy can be captured with a turbine, which is connected to an electric generator. The technology of wind turbines is well developed, and the cost of wind-generated electricity is lower than that produced by some other sources. The primary disadvantage of wind energy is that few regions have winds strong or consistent enough to make wind turbines economical. In addition, large, commercial wind turbines may not be attractive additions to the landscape. At full speed they can be very noisy. The blades of large wind turbines can also interfere with microwave communications. While wind energy may never be a major source of energy, it is a practical energy alternative in some areas.

Exploration 6 Teacher's Notes, continued

Teaching Strategies

In order to complete this Exploration successfully, students must have a clear understanding of the necessary mathematical calculations. These calculations are outlined in the CD-ROM articles. You may wish to review the CD-ROM articles with students who are having difficulty determining the correct procedure. Make sure students understand that they must compare the total energy savings over a long period of time with the cost of installing the wind turbine. If the wind turbine pays for itself before it needs repairs, then the turbine can be considered cost-effective.

As an extension of this Exploration, you may wish to invite an energy-resource expert from a local college, university, or scientific institution to speak to the class about current energy-efficient technologies.

Bibliography for Teachers

Brower, Michael. *Cool Energy: Renewable Solutions to Environmental Problems.* Revised edition. Cambridge, MA: MIT Press, 1992.

Burmeister, George, and Frank Kreith. *Energy Management and Conservation.* Washington, DC: National Conference of State Legislatures, 1993.

Bibliography for Students

Cozic, Charles, ed. *Current Controversies: Energy Alternatives.* San Diego, CA: Greenhaven Press, 1994.

Inhaber, Herbert, and Harry Saunders. "Road to Nowhere," *The Sciences,* Nov.–Dec. 1994, pp. 20–25.

Other Media

Electric Bill
Software (Apple II family)
Queue, Inc.
338 Commerce Dr.
Fairfield, CT 06432
203-335-0906
800-232-2224

Power Struggle
Videotape
Bullfrog Films
P. O. Box 149
Oley, PA 19547
215-779-8226
800-543-3764

Interested students may also find relevant information about alternative energy sources on the Internet. Suggest that students use keywords such as the following to conduct their search: *energy, alternative sources of energy, fossil fuels, conservation, electricity,* and *wind turbines.* Students may also be interested in searching for information about *energy-efficient homes.*

EXPLORATION 6 • THE GENERATION GAP 49

Name _____ Date _____ Class _____

Exploration 6
Worksheet

The Generation Gap

1. Wendy Powers is a home builder who is considering a plan to make her homes more efficient. What has she asked you to do to help her?

   _____
   _____
   _____
   _____

2. Dr. Labcoat has set up a system that enables you to test the energy output of the wind turbine at eight different speed settings. Run the tests, and record your results below.

Meters per second	Kilowatt-hours	Time-lapse indicator

3. What is the value of the above information?

   _____
   _____
   _____
   _____
   _____

Name _____ Date _____ Class _____

Exploration 6 Worksheet, continued

4. What other information will you need to complete your task?

   _____

   _____

   _____

   _____

   _____

   _____

   _____

5. Use the lab resources to find this information. You can record your notes here.

   _____

   _____

   _____

   _____

   _____

   _____

   _____

   _____

   _____

   _____

   _____

   _____

   _____

   _____

6. How will you calculate the amount of money a wind turbine can save a homeowner over the course of a year?

   _____

   _____

   _____

   _____

   _____

EXPLORATION 6 • THE GENERATION GAP

Name _____ Date _____ Class _____

Exploration 6 Worksheet, continued

7. Use the table below to record the energy output of the Electroprop wind turbine and your calculations of the savings it will bring the homeowner.

Wind speed in meters per second (m/s)	Energy output over 7 days in kilowatt-hours (kWh)	Savings per year ($)	Years until Electroprop wind turbine has paid for itself

Record your conclusions in the fax to Ms. Powers.

Name _____ Date _____ Class _____

Exploration 6
Fax Form

FAX

To: Ms. Wendy Powers (FAX 415-555-2766)

From:

Date:

Subject: Wind-Energy Economics

Is it cost-effective to use the Electroprop to generate energy in the San Francisco area? Why or why not?

_____

_____

_____

_____

✂··

For Internal Use Only

Please answer the following questions for my laboratory records. Scientists must always keep good records. *Dr. Crystal Labcoat*

Approximately how much money would the Electroprop save a San Francisco homeowner in an average year?

$4	$28	$140	$400	$1460	$2800

Approximately how many years would it take for the Electroprop to pay for itself?

1	5	10	16	28	250

How many years would it take for the Electroprop to pay for itself if the average wind speed in the San Francisco area were each of the following:

8 m/s? _____

5 m/s? _____

2 m/s? _____

EXPLORATION 6 • THE GENERATION GAP 53

The Generation Gap

The following articles can also be found by accessing the computer graphic of the CD-ROM for Exploration 6:

- The Cost of Electricity
- Wind Turbines

Exploration 6
CD-ROM Articles

```
ELECTRIC BILL
IN 29 DAYS YOU USED        249 KWH
  READ DATE              METER #0000
  03/29                        5121
  02/29                        4872
  DIFFERENCE                249 KWH
  RATE CALCULATION:         249 KWH
  RESIDENTIAL SERVICE
  CUSTOMER CHARGE:            $6.00
    ENERGY:   $.03550/KWH     $8.84
    FUEL:     $.01583/KWH      3.84
  SUBTOTAL ELECTRIC CHARGES  $18.78
    SALES TAX                   .19
  TOTAL COST FOR
  ELECTRIC SERVICE:          $18.97
```

A residential customer pays for each kilowatt-hour of electrical energy.

The Cost of Electricity

Calculating Electricity Costs

Electricity usage is measured by a unit called the *kilowatt-hour*, abbreviated *kWh*. The average American household uses approximately 1000 kWh of electricity each month. Utility companies charge each household for the electricity used, and the cost of electricity varies from city to city. Consider the cost of electricity in five American cities.

Average Monthly Electricity Usage and Cost

City	Cost per kWh
Atlanta, GA	$0.06
Chicago, IL	$0.05
Dallas, TX	$0.05
Houston, TX	$0.09
San Francisco, CA	$0.14

What is the average cost of electricity per month for a household in each city? To find out, multiply the number of kilowatt-hours of electricity used by the cost per kilowatt-hour. For example, the cost of electricity in Atlanta is six cents per kilowatt-hour. Since the average household uses 1000 kWh of electricity in one month, the average cost of electricity per month is:

1000 kWh × $0.06 = $60.00

How to Read an Electric Bill

The charge for your electricity may be included in a bill along with charges for water, garbage collection, and other services. Or charges for electricity may come in a separate bill. Either way, the bill will show how much electricity was used in a given amount of time and the cost of the electricity.

Consider the following example of an electric bill.

```
IN 29 DAYS YOU USED              831 kWh
READ DATE              METER # 00328450
05/07/96                            1463
04/08/96                             632
DIFFERENCE                           831

RATE CALCULATION
CUSTOMER CHARGE                    $6.00
831 kWh AT $0.09/kWh               74.79

SUBTOTAL ELECTRIC CHARGES         $80.79

SALES TAX                           0.81

TOTAL COST FOR
ELECTRIC SERVICE                  $81.60

FOR THIS 29-DAY PERIOD, YOUR
AVERAGE DAILY COST FOR
ELECTRIC SERVICE WAS               $2.81
```

54 HOLT SCIENCE AND TECHNOLOGY INTERACTIVE EXPLORATIONS TEACHER'S GUIDE

The first part of the bill shows you that on April 8, the meter read 632 kWh. On May 7, the meter read 1463 kWh. The difference, 831 kWh, is the total number of kWh used from April 8 to May 7.

The next part of this bill shows the actual charges. First, it shows a service charge of $6.00. This charge covers the cost of reading the meter and processing the bill. The cost of the electricity follows. The number of kilowatt-hours is multiplied by the city's charge per kilowatt-hour—nine cents in this example. The service charge and the electricity charge are subtotaled, and a sales tax is computed.

The last part of this bill shows the total cost for electric service and the average cost per day for electricity.

In this sample electric bill, the charge per kilowatt-hour represents the cost of the actual fuel used to produce the electricity and the cost of operations and maintenance at the utility company. In some cities, these charges may be represented separately. In addition, some utility companies may charge an increased rate if a household uses more than a certain number of kilowatt-hours, or the charge per kilowatt hour may vary depending on what time of day the electricity is used.

How to Lower Your Household's Electric Bill

Electric bills can cost your household a large amount of money each month. However, you can do several things to help reduce the amount of electricity used and thus lower your household's total energy bill.

Use less hot water. In many homes, water is heated using electricity. Keep showers short, and use a cold-water rinse cycle when using the washing machine.

Take charge of the thermostat. Electricity is often used to heat and cool homes. In the winter, lower the thermostat and wear a sweater indoors. In the summer, raise the thermostat and use fans to keep cool.

Turn off the lights. Leaving the lights on when they are not in use wastes electricity. Lights also generate heat and can add to cooling costs in the summer.

Turn off appliances. Turn off the TV, radio, and other appliances when no one is using them.

Wind Turbines

What Is Wind?

The movement of air across the Earth's surface is known as *wind*. Winds are set in motion by the uneven heating of the Earth's surface, and they tend to blow from regions of higher to lower air pressure. Winds are divided into two types—*local* and *planetary*.

Consider an oceanside setting. During the day, the land heats up faster than the water, and the warm land heats the air above it. As the hot air rises, the pressure it puts on the land decreases. The cool air over the water is heavy and dense, thus creating an area of high pressure. This cool, heavy air over the water then moves to the low-pressure area over the land. At night, the reverse occurs. Land cools faster than water. Thus, the air moves from the high-pressure area over the land to the low-pressure area over the ocean. Winds created by this process are known as *local winds*. Local winds are set in motion by the surface features of a particular area, such as bodies of water, ice formations, and land formations.

Pressure differences across the globe also create winds called *planetary winds*. Air rises in warm regions along the equator, creating areas of low air pressure. After a time this air stops rising and drifts toward the polar regions. Most of this air cools and sinks before it reaches the poles, creating areas of high pressure. The pattern of rising and sinking air creates pressure belts along the Earth's surface. Planetary winds flow between pressure belts, from areas of high pressure to areas of low pressure.

Exploration 6 CD-ROM articles, continued

What Is a Wind Turbine?
A wind turbine is a device that harnesses the energy of wind and converts it into electrical energy. In a wind turbine, wind turns large propellers that are attached to a generator. As the generator spins, it produces electricity that can be carried along power lines.

A wind turbine uses renewable wind energy to make electricity.

A typical wind turbine has a tower, blades, a generator and cables, and an electronic control system. The tower raises the blades to a point where the wind is strong and blows steadily. A typical turbine has two or three blades that span about 18 m. The generator produces electricity from the spinning blades, and the electricity travels down the tower along the cables. At the base of the turbine, a control system monitors the electricity production.

A wind power plant, or wind farm, has large clusters of wind turbines. These are often owned by private businesses that sell the electricity produced to utility companies. Because wind turbines are large, a successful wind farm requires a large amount of land along with appropriate wind conditions.

Advantages and Disadvantages of Using Wind as an Energy Source
There are several advantages to using wind as an energy source. First and foremost, wind is a renewable resource because it is produced by the sun. The sun heats the Earth's surface unevenly and causes disturbances in the atmosphere, forming winds. As long as the sun shines, the Earth will have winds. Another advantage of wind turbines is that they can use the wind without adding pollutants to the air, water, or soil.

Also, wind-generated electricity is often less expensive than that produced by other sources. Modern wind farms can produce electricity at just five cents per kilowatt-hour. Traditional power plants produce electricity at five to six cents per kilowatt-hour, require more expensive maintenance, and create more pollution.

Harnessing wind energy also has a number of disadvantages. To be effective, wind turbines require steady winds that do not fall below 3.1 m/s. The most efficient production of energy occurs when wind speed averages about 4.5 m/s. Many regions lack winds that are consistent and strong enough to make a wind turbine system affordable. In addition, most wind farms can only produce electricity about 25 percent of the time; traditional power plants, such as coal-operated plants, can operate around the clock. Also, commercial wind generators can be noisy, and their large propellers can make the landscape unattractive.

An anemometer measures wind speed.

Wind Speeds in the United States
The chart below shows the average wind speed in seven United States cities. As you can see, average wind speed varies around the nation.

City	Wind speed (m/s)
Boston, MA	5.0
Las Vegas, NV	4.0
New Orleans, LA	3.0
Oklahoma City, OK	6.0
Richmond, VA	3.0
San Francisco, CA	6.0
St. Louis, MO	4.0

Most wind turbines operate in the Midwest, the Northeast, and California. In fact, wind energy produces about 1 percent of California's electricity, making California the leading state for harnessing wind resources.

Since wind speed varies, homes that use wind power for their electricity must also have a second energy source. On days when the wind speed is insufficient, the second energy source provides electricity. When the wind turbine produces more electricity than the household needs, the excess can be sold to a utility company. Sometimes the sale of this excess energy can save a household 50 to 90 percent on its monthly electric bill.

A Wind Farm

Wind Turbines—Are They Worth It?

Wind turbines can be a safe and affordable method of generating electricity. To determine whether a wind turbine will be an energy source for a household, the following factors must be considered:

- average wind speed in the region
- number of kilowatt-hours produced by the wind turbine per week
- cost of kilowatt-hours charged by the local utility company
- cost of wind turbine

Here is one example of a household that found that money could be saved by using a wind turbine. In Nevada, where the Hioes live, the average wind speed is 4.0 m/s, which produces about 65 kWh of electricity in one week. The local utility company charges 13 cents per kWh. On average, the Hioes use 1000 kWh per month, or 12,000 kWh per year.

Approximate yearly savings using a wind turbine:

Electricity from wind turbine:
    65 kWh × 52 weeks = 3380 kWh per year

Average cost of electricity from the local utility company:
    $0.13 per kWh

Savings per year:
    3380 kWh × $0.13 = $439.40

Using one wind turbine to generate electricity, the Hioes could save approximately $439.40 in one year.

How long would it take for the wind turbine to pay for itself?

If the initial cost of the Hioes' wind turbine is $7000, it would take approximately 16 years for the wind turbine to pay for itself.

Exploration 7
Teacher's Notes

Teach It While It's Hot!

Key Concepts	Heat and temperature are different concepts. Graphs can be used as a visual aid to make these and other scientific concepts more readily understandable.
Summary	In the past, Mr. Kelvin McCool, a science teacher at J. P. Joule Middle School, has experienced some problems teaching his students the difference between temperature and heat. He has asked Dr. Labcoat and her assistants to help him by preparing a demonstration and some graphs to illustrate the characteristics of both temperature and heat.
Mission	Help Mr. McCool teach the relationship between heat and temperature to his class of middle-school students.
Solution	Heat refers to the total energy of motion of the atoms and molecules in a substance, while temperature is a measure of the average energy of motion of the atoms and molecules in a substance. A volume of water twice the size of another volume of water must contain twice as much heat to maintain the same temperature as the smaller volume.
Background	Temperature is an average measure of the movement of the constantly vibrating particles that make up all matter. The hotter a substance gets, the more energy its particles have, and the faster the particles move. The colder a substance gets, the less energy its particles have, and the slower the particles move. Scientists theorize that all movement of particles stops at absolute zero (–273°C). By using laser devices in laboratories, scientists have cooled matter to within a millionth of a degree of absolute zero.
	Working with extremely cold temperatures has resulted in the development of some important technologies. Cryosurgery allows doctors to use low temperatures to seal off blood vessels during an operation. Superconductors are materials that have been cooled to super-low temperatures so that they can conduct electricity without resistance. The opportunities made possible by a perfectly efficient conductor are nearly endless.
Teaching Strategies	Successfully completing this Exploration depends on thoroughly researching the CD-ROM articles and correctly interpreting the graph of the results. Students must recognize that the amount of heat needed to raise the temperature of a mass of water a specific amount is proportional to that mass. Because the density of water = 1 g/1 mL, this proportion can also be discussed in terms of volume.

Exploration 7 Teacher's Notes, continued

For instance, a 300 mL volume of water will require three times as much heat as a 100 mL volume to raise its temperature the same amount. You can use the graph in the Exploration to emphasize the differences in the amount of heat required to raise different volumes of water to specific temperatures. Encourage students to study the graph of their results carefully before attempting to complete the fax form. You may wish to review the CD-ROM articles on graphing with students who are having difficulty reading the graph.

As an extension of this Exploration, have students use what they discovered to create a graph that shows the amount of heat required to raise the temperature of 400 mL, 500 mL, and 600 mL of water from 20°C to 100°C. You may also wish to suggest that students discuss the advantages of understanding the relationship between heat and temperature. For instance, ask students how a cook might change his or her approach to heating food based on a knowledge of this relationship. A larger (more massive) pan or pot would require more heat energy to heat up than one that is less massive. For example, it would take longer to boil a certain volume of water in a 2 kg pan than in a 1 kg pan.

Bibliography for Teachers

Cuevas, Mapi M., and William G. Lamb. *Physical Science*. Austin, TX: Holt, Rinehart and Winston, 1994.

Bibliography for Students

Maury, Jean-Pierre. *Heat & Cold*. Hauppauge, NY: Barron's Educational Series, Inc., 1989.

Wood, Robert W. *Physics for Kids—Forty-Nine Easy Experiments with Heat*. Blue Ridge Summit, PA: TAB Books Inc., 1989.

Other Media

Heat and Energy Transfer
Film and videotape
Coronet/MTI
P. O. Box 2649
Columbus, OH 43216
800-777-8100

Heat: Molecules in Motion
Videodisc
AIMS Media
9710 DeSoto Ave.
Chatsworth, CA 91311-4409
818-773-4300

Heat, Temperature, and the Properties of Matter
Film and Videotape
Coronet/MTI
P. O. Box 2649
Columbus, OH 43216
800-777-8100

In addition to the above resources, interested students may find relevant information about heat and temperature on the Internet. Suggest that students conduct a search with keywords such as the following: *heat and temperature, heat and energy, Newtonian universe,* and *James Joule.*

Name _____ Date _____ Class _____

Exploration 7
Worksheet

Teach It While It's Hot!

1. What has Dr. Labcoat asked you to do to help Mr. McCool?

2. What information would be helpful to know before you begin your investigation?

3. Where do you think you could find this information?

4. What happens when heat energy is applied to a beaker of water?

5. Record your observations as each beaker (quantity) of water is placed on the ring stand.
   a. green (100 mL)

   b. red (200 mL)

Exploration 7 Worksheet, continued

c. blue (300 mL)

6. How can you calculate the amount of heat required to increase the temperature of 600 mL of water from 20°C to 100°C?

7. Why did Dr. Labcoat provide you with three different quantities of water?

Name _____ Date _____ Class _____

Exploration 7 Worksheet, continued

8. Use the graph as well as your knowledge of temperature and heat to describe what this demonstration shows.

   _____

   _____

   _____

   _____

   _____

   _____

   _____

   _____

   _____

   _____

   _____

   _____

   _____

   _____

9. Based on what you've learned during this activity, would you recommend this demonstration to Mr. McCool? Why or why not?

   _____

   _____

   _____

   _____

   _____

   _____

   _____

Record your conclusions in the fax to Mr. McCool.

Name _____ Date _____ Class _____

Exploration 7
Fax Form

FAX

To: Mr. Kelvin McCool (FAX 512-555-4328)

From:

Date:

Subject: Teaching Recommendations

What relationship is represented by your graph?

_____

_____

_____

_____

_____

_____

_____

Please use your data to determine the answers to the following questions:

Which beaker contains the most heat energy at 100°C?	GREEN	RED	BLUE
Approximately how much heat would have to be added to increase the temperature of 600 mL of water from 20°C to 100°C?	100,000 joules	200,000 joules	300,000 joules

What is the approximate temperature of each sample of water when the amount of heat energy added is 30,000 joules?

GREEN: RED: BLUE:

EXPLORATION 7 • TEACH IT WHILE IT'S HOT!   63

Name _____ Date _____ Class _____

Exploration 7 Fax Form, continued

Please write and answer one essay question that will help Mr. McCool's students understand the relationship between temperature and heat.

Teach It While It's Hot!

The following articles can also be found by accessing the computer graphic of the CD-ROM for Exploration 7:

- *Temperature and Heat*
- *Graphs*

Temperature and Heat

What's the Difference?

Compare the following two definitions:

Temperature—the measure of the average kinetic energy of the motion of atoms and molecules in a substance

Heat—the internal energy of a substance that is due to the random movement of the atoms and molecules in a substance

If you compare these definitions closely, you will see that temperature refers to the *average kinetic energy* of a substance while heat refers to the *total kinetic energy* of a substance. To explore this difference further, imagine that you have two mugs of hot cocoa, both at the same temperature and both made of the same material. Would you say that they both contain the same amount of heat energy? The answer is yes if they both have the same mass.

But what if one mug were larger and had more mass than the other mug? Would the more massive mug have the same amount of heat as the smaller one, even if they were both at the same temperature? The answer is no. Even though the molecules in each mug have approximately the same kinetic energy, the larger mug has more molecules, and therefore has more heat energy.

The larger mug has more cocoa and more heat.

Mass Is Related to Heat

Now think about making some eggs and potatoes to go with the cocoa. First you must heat two frying pans made of the same material. Let's say that one pan has a mass of 1 kg and the other pan has a mass of 2 kg. If you placed the two pans on identical burners set to the same temperature, the heavier frying pan would take twice as long to get hot. That's because the more massive the object, the more heat energy it takes to raise its temperature.

Suppose that you wanted to raise the temperature of each frying pan by 1°C. To do this, the 2 kg frying pan would require exactly twice as much heat as the 1 kg frying pan would require. So the effect of mass on temperature can be summed up this way: when the mass of a substance is doubled, it takes twice as much heat to increase the temperature of that mass by 1°C.

The larger frying pan takes longer to get hot.

Measuring Heat

In the metric system, heat energy is measured in *joules*. The joule is named after James Prescott Joule, a nineteenth-century scientist who theorized that mechanical energy and heat energy were different forms of the same thing. A joule (J) is approximately equal to the amount of mechanical energy required to raise a 100-gram mass 1 meter. If measuring heat energy, it takes 4.19 J of heat to raise the temperature of 1 g of

EXPLORATION 7 • TEACH IT WHILE IT'S HOT! **65**

water by 1°C. Heat can also be measured in calories—one calorie equals the amount of heat needed to increase the temperature of 1 g of water by 1°C. So 1 calorie = 4.19 J.

Measuring Temperature

Thermometers are used to measure temperature. Many thermometers work on the principle that matter expands when heated and contracts when cooled. A glass thermometer, for example, has a narrow tube that is filled with either mercury or alcohol. These liquids are used because they do not freeze or boil easily. When heated, the mercury or alcohol expands inside the tube. And since the rate of expansion for these liquids is predictable, the glass tube can be calibrated with degree marks that correspond to a temperature scale.

The temperature scales most often used are the Celsius and Fahrenheit scales. In the metric system, water freezes at 0°Celsius (°C) and boils at 100°C. On the Fahrenheit scale, water freezes at 32°Fahrenheit (°F) and boils at 212°F. The Fahrenheit scale is named after the German physicist Gabriel Fahrenheit, who in 1714 invented the glass mercury thermometer.

Specific Heat

Raising a substance's temperature always means increasing the heat energy of the substance; it always takes energy to raise temperature. But some substances need less energy to heat up. The term *specific heat* describes the amount of heat needed to raise the temperature of 1 g of a substance by 1°C. If you have 1 g of water, you have to add 1 calorie of heat to increase the temperature by 1°C. But most substances need a lot less energy to get hot. For instance, you need only 0.22 calories of heat to raise the temperature of 1 g of aluminum by 1°C.

Few substances have a specific heat as great as that of water. Water can absorb a large amount of heat energy without a large change in its temperature. In fact, the ocean (which feels cool when we go swimming in it) actually holds an enormous amount of heat energy. Its large mass combined with the high specific heat of water keep the ocean temperature relatively cool.

Graphs

Types of Graphs

Often scientists use various types of graphs to illustrate information. Check out the following three examples of graphs. They all show information that would not be as clear if described by words only.

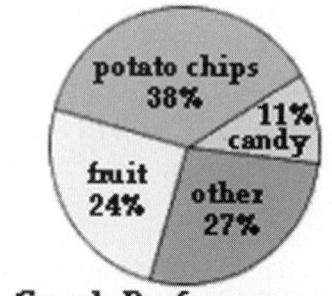

Snack Preferences of Middle School Students

potato chips 38%
candy 11%
other 27%
fruit 24%

Exploration 7 CD-ROM articles, continued

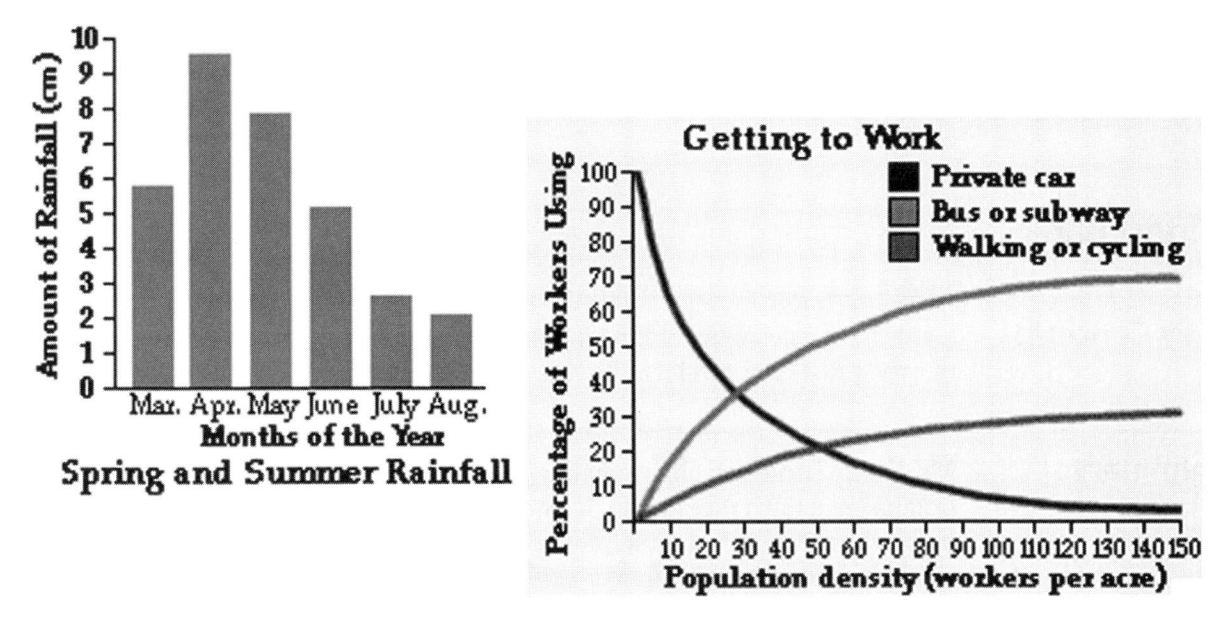

Parts of a Graph

The diagram below shows the parts of a graph. Note that both the *x*- and *y*-axes increase in regular intervals. That makes reading the graph easier. A good graph always has clear labels and a title.

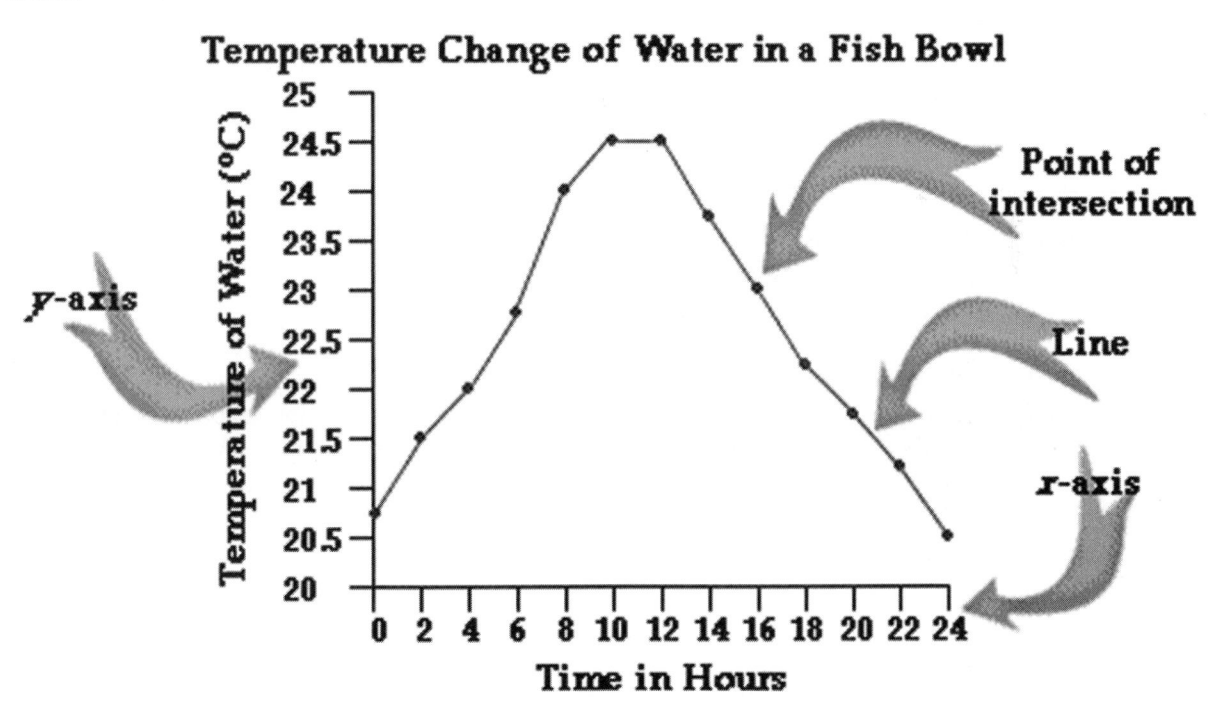

EXPLORATION 7 • TEACH IT WHILE IT'S HOT! 67

Exploration 8
Teacher's Notes

Flood Bank

Key Concepts	Dams have a powerful effect on natural river environments. Regulated flooding of a river does not produce the same results as natural flooding.
Summary	Ms. Sandy Banks, as chairman of her local environmental-impact committee, is organizing a town meeting to debate whether a dam should be built in the community to create a reservoir. She needs to know what long-term effects a dam would have on the natural river environment.
Mission	Predict the effects of a dam on the natural flow of a river, its geological formations, and its ecosystem.
Solution	Dams block the natural flow of a river and limit the amount of sediment that the river carries downstream. This lack of sediment can drastically affect ecosystems that depend on the sediment in the river as the basis of their food chains. Additionally, regulated flooding does not produce the same scouring of the river bottom that natural flooding does. The resulting changes in the landscape threaten the organisms that inhabit the river.
Background	Altering the flow of river water can change the ecosystem of not only the local river environment but also the environment many kilometers downstream. Large, fan-shaped deposits at the mouth of a river are called deltas, and they are home to a wide variety of fish, crustaceans, and mollusks. Because the ocean is continually sweeping portions of the sediment out to sea, a delta depends on new large deposits of sediment from the river each year. The Mississippi River, for example, deposits 2 million tons of sediment into the Gulf of Mexico each year. A dam upstream would limit the amount of sediment that the river would be able to carry downstream, and the delta and its ecosystem could potentially disappear.

Exploration 8 Teacher's Notes, continued

Teaching Strategies	Successfully completing this Exploration requires that students research the CD-ROM articles thoroughly. Encourage students to use the Notepad function to take notes from the articles.

As an extension of this Exploration, you may wish to have students research natural alterations to the courses of rivers and streams, such as dams built by beavers. Ask students to find out which plants and animals depend on these natural water regulators and how removing beaver dams can disrupt ecosystems by increasing the rate of the river's flow. |
| **Bibliography for Teachers** | "Science and Technology: The Beautiful and the Dammed." *The Economist,* 322 (7752): March 28, 1992, p. 93.

Sides, Hampton. "Let There Be High Water." *Outside,* 21 (7): July 1996, pp. 38–41, 104–106. |
| **Bibliography for Students** | Kaufman, Jeffrey S., Robert C. Knott, and Lincoln Bergman. *River Cutters.* Berkeley, CA: Lawrence Hall of Science, 1990.

Wuerthner, George. "Dammed River, Doomed Mollusks." *Defenders,* 67 (3): May/June 1992, pp. 12–13. |
| **Other Media** | *Erosion and Weathering: Looking at the Land*
Erosion: Leveling the Land
　　Two videotapes
　　Britannica
　　310 S. Michigan Ave.
　　Chicago, IL 60604-9839
　　800-554-9862

Interested students may find relevant information about dams and their effects on the environment by accessing the Internet. Suggest that students conduct a search using keywords such as the following: *river dams, flooding,* and *altering ecosystems.* |

EXPLORATION 8 • FLOOD BANK **69**

Name _____ Date _____ Class _____

Exploration 8 Worksheet

Flood Bank

1. Ms. Sandy Banks is the chairperson of her local environmental-impact committee. What has she asked you to do to help her?

   _____
   _____
   _____
   _____

2. How can the simulation that Dr. Labcoat has provided help you with your investigation?

   _____
   _____
   _____
   _____
   _____
   _____

3. Conduct your research, recording your notes in the space provided below.

   _____
   _____
   _____
   _____
   _____
   _____
   _____
   _____
   _____
   _____
   _____

Name _____ Date _____ Class _____

Exploration 8 Worksheet, continued

4. Use the table below to record the effects of both types of water flow at varying flow rates.

Type of flow	Low	Medium	High
Natural flow			
Regulated flow			

5. Compare the results of your stream-table observations with the information you discovered in your research. Which environmental and geological effects of dams are not reproducible by a stream table?

_____

_____

_____

_____

_____

_____

Record your conclusions in the fax to Ms. Banks.

EXPLORATION 8 • FLOOD BANK

Name _____ Date _____ Class _____

Exploration 8
Fax Form

FAX

To: Ms. Sandy Banks (FAX 520-555-7239)

From:

Date:

Subject: Possible effects of a dam on the natural river environment

What effect does a dam have on the natural river environment?

Name _____ Date _____ Class _____

Exploration 8 Fax Form, continued

Which volume of water flow has the greatest impact on the formation of these geologic features?

| Low | Medium | High |

Is it possible to maintain a healthy river environment downstream from a dam?

| YES | NO |

Why or why not?

_____
_____
_____
_____
_____
_____
_____
_____
_____
_____
_____
_____
_____
_____
_____
_____
_____
_____
_____
_____
_____

Exploration 8
CD-ROM Articles

Flood Bank

The following articles can also be found by accessing the computer graphic of the CD-ROM for Exploration 8:

- Anatomy of a River
- Questions About Controlling a River
- A Case Study

Anatomy of a River

From Top to Bottom

Rivers flow from areas of higher elevation to areas of lower elevation. Several rivers get their start in the high elevations of the Rocky Mountains. Rainwater and snowmelt run down the slopes of the mountains, forming small channels that eventually become streams and rivers. These streams and rivers empty into larger rivers such as the Colorado River, which runs west from the Rocky Mountains. The rivers and streams that empty into larger rivers are called *tributaries*. For example, the Illinois River is a tributary of the mighty Mississippi River. Most river systems will continue to flow downhill until their water reaches the sea. The point where a river empties into the sea is called the river's *mouth*.

As the water of a river flows, some of the land along the river and along the river bottom is washed downstream. As the flowing water erodes the riverbanks and river bottom, it becomes rich in *sediment* consisting of small rocks, soil, and other particles. The sediment carried by the river is often called the river's *load*. The load of a river increases with the river's size and flow rate; a large, swift river causes more erosion and carries more sediment than a small, slow-moving river.

A river's load is not always constant. For example, a river will deposit some of its load wherever the river water slows down—such as on the inside of curves, in areas where the river widens, or where the riverbanks are less steep. This process is called *deposition*.

Growth and Aging

Rivers change over time. Compare the differences between a young, mature, and old river.

Young river: A young river is characterized by steep banks, narrow or V-shaped valleys, and rapid erosion. Young rivers often form steep gorges or canyons.

Mature river: A mature river is characterized by gently sloping banks, wide valleys, and moderate erosion.

Old river: An old river is characterized by low banks, wide and flat valleys, and little erosion. Wide curves called *meanders* mark the shape of an old river. And although an old river does not cause much erosion, it does carry and distribute large amounts of sediment that it receives from its tributaries.

Some rivers can be young at their source, mature in the middle, and old near their mouth. As the river ages, it spreads over the *flood plain*. The flood plain is the flat, low-lying area that is covered by flowing water during times of high water volume.

Seasonal Flow and Flooding

The flow rate of a river can change from season to season and even from day to day. In the spring, for example, mountain streams swell with water as the snow melts. This causes a surge in the river's flow rate, even far downstream.

Floods occur when heavy rain or melting snow enters a river system causing the river to overflow its banks. A flood causes a tremendous increase in erosion. A flood can move boulders as well as tons of sediment. A flood can even change the course of a river.

When the flood water slows down, it deposits sediment along the river bank and the flood plain. This sediment is very rich in nutrients. And when it is deposited in the flood plain, it serves to maintain healthy sandbars, which in turn provide spawning grounds for many species of fish and other organisms.

Even though floods appear to be very damaging to the environment, they also serve to maintain the natural ecosystems in and around the river. On the other hand, human-made structures on a flood plain are very much at risk during floods. Because floods are so destructive to anything in their path, people try to control the flow of the water by deepening the riverbeds, by building walls along the riverbanks, and by constructing dams.

Questions About Controlling a River

What Does a Dam Do?

Dams enable us to control the flow of water in a river. A dam built across a river creates an artificial lake, called a *reservoir,* behind the dam. Water can be released from the reservoir at a rate that prevents flooding downstream. Water held in the reservoir is often used for drinking, for recreational purposes, and for irrigating farms.

In addition to the uses listed above, some dams are equipped with electrical generators that operate as the water flows over large turbines. These dams generate electricity and are often referred to as hydroelectric dams.

How Does a Dam Affect the River's Environment?

Because a dam blocks the natural flow of a river, it limits the erosion and deposition of sediment by the river. So instead of collecting and carrying nutrient-rich sediment downstream, the river drops the sediment at the bottom of the reservoir. As a result, the ecosystems downstream are deprived of the sediment that is often the basis of their food chains.

The water released through a dam is also cooler than naturally flowing river water. Many organisms cannot adapt to this cooler water, and die off. Again, this may break a natural food chain and cause serious damage to the river ecosystem.

Dams also prevent flood waters from periodically scouring the river bottom. As a result, debris settles on the river's bottom and creates barriers in front of inlets where fish would normally spawn. Where flood waters once carved out the river's banks and created new sandbanks farther downstream, vegetation takes root. These changes in the landscape threaten the organisms that live in and along the river system.

Can We Control the Flow to Protect the River System?

Dams are very disruptive to the natural river environment. Even if water could flow through the dam at the same rate it would normally flow, much of the river's sediment would remain on the bottom of the reservoir. Controlled flooding could help keep the river clear of debris and could help distribute some of the sediment below the dam, but sediment above the dam would still need to be moved somehow from behind the dam to areas downstream. Also, some method would need to be devised to increase the temperature of the water before it affected the ecosystems downstream.

A Case Study

The Colorado River and the Glen Canyon Dam

Before construction of the Glen Canyon Dam in 1963, spring floods of over 93,000 cubic feet per second rushed down Glen Canyon, eventually flooding through the Grand Canyon. However, since 1963, these spring floods of the Colorado River have been controlled by Glen Canyon Dam.

The spring floods once cleaned the riverbed, built beaches along the banks, and removed vegetation that clogged the river's path. Since 1963, however, vegetation has thickened, nutrient levels have declined, sediment has been lost, water temperatures have dropped, and natural sandbars have disappeared. Many of the species that lived in the Colorado River before the dam was built, such as the humpbacked chub, are now endangered. Others are extinct. New species, such as the rainbow trout, have been introduced to the river and have disrupted the ecosystem's food web.

After more than 10 years of data collection and planning, scientists and government officials began an experiment. On March 27, 1996, a controlled flood was released from Glen Canyon Dam. Each day for one week, 45,000 cubic feet per second of water cascaded from the dam. It was hoped that this controlled flooding would restore some of the ecological conditions that existed in the Colorado River before the dam was built. Early results indicate that although the natural sediment is still being held back by the dam, some of the beaches and sandbars and some of the natural habitats have been temporarily restored.

Exploration 1
Teacher's Notes

What's Bugging You?

Key Concepts	Humans can have commensalistic, mutualistic, and parasitic interactions with organisms that live in and on the human body. Parasitic interactions can cause human illnesses.
Summary	Several of Dr. Mike Roe's patients have the same symptoms of illness. Dr. Roe has sent samples of six organisms that he collected from each of his patients. He wants to know the identity of these samples as well as the type of interaction each organism has with its human host. If the organism that is making his patients sick is a parasite, Dr. Roe would like to know how it was transmitted to its host.
Mission	Describe the effects of the sample organisms on humans, and identify the organism that is making Dr. Roe's patients sick.
Solution	Samples 1 and 3 are commensalistic, sample 5 is mutualistic, and samples 2, 4, and 6 are parasitic. Organisms 1 through 6, respectively, are *Entamoeba gingivalis, Enterobius vermicularis, Demodex folliculorum, Rickettsia rickettsii, Propionibacterium acnes,* and *Streptococcus mutans.* Sample 4, *Rickettsia rickettsii,* is the organism that is making Dr. Roe's patients sick. This parasitic bacterium lives in ticks and is transmitted when infected ticks bite humans.
Background	Not all of the organisms that live in or on the human body are parasites. For example, *Lactobacillus acidophilus* lives in milk, yogurt, and other dairy products and helps to maintain healthy levels of helpful bacteria in the human body. The strain of *Escherichia coli* that lives in the digestive tracts of humans and other animals has a mutualistic relationship with its hosts. It plays an important role in digestive processes. However, unfamiliar strains of *E. coli,* such as those found in undercooked meats, can cause illnesses in humans.
	Scientists and medical practitioners identify and analyze organisms found in or on the human body so that they can better diagnose diseases and prevent illnesses. At pharmaceutical laboratories, researchers study the effects of natural and synthetic antibiotics on disease-causing bacteria. Natural and holistic medical practitioners study the commensalistic and mutualistic interactions between humans and organisms that live in or on the human body in order to prevent sickness and promote overall wellness.

Exploration 1 Teacher's Notes, continued

Teaching Strategies

Students must make numerous observations in the lab before successfully completing this Exploration. To increase students' efficiency, emphasize that they should look at the reference slides before looking at the sample slides. This will familiarize the students with certain characteristics of the reference organisms so that when they do examine the sample slides, it will be easier for them to identify each sample organism. Remind students that the colors of the reference and sample slides may not be indicative of each organism's classification. The colors are simply the stains used to make the organisms easier to see under the microscope. Once the sample organisms have been identified, encourage students to research the CD-ROM articles thoroughly to identify the interactions the organisms have with their hosts.

As an extension of this Exploration, you may wish to have students explore other kinds of organisms that have parasitic, mutualistic, or commensalistic relationships with humans and animals. Students could present their findings in a simple interaction diagram.

Bibliography for Teachers

Davis, George M. "Systematics and Public Health." *Bioscience,* 459 (10): November 1995, pp. 705–714.

Levins, Richard, Tamara Awerbuch, Uwe Brinkman, Irina Eckhardt, and Paul Epstein. "The Emergence of New Diseases." *American Scientist,* 82 (1): January/February 1994, pp. 52–60.

Bibliography for Students

Lemonick, Michael D. "The Killers All Around." *Time,* 144 (11): September 12, 1994, pp. 62–69.

"Home Remedies: From Soup to Oats." *University of California at Berkeley Wellness Letter,* 12 (2): November 1995, pp. 6–7.

Other Media

Bacteria and Health Videodisc
SVE (Society for Visual Education)
6677 N. Northwest Highway
Chicago, IL 60631
800-829-1900

In addition to the above videodisc, there are many resources on the Internet that explore the relationships between humans and other organisms. Interested students can access relevant articles with keywords such as the following: *commensalism, mutualism, parasitism, microorganisms, bacteriology, wellness,* and *antibiotics.*

78 HOLT SCIENCE AND TECHNOLOGY INTERACTIVE EXPLORATIONS TEACHER'S GUIDE

Name _____ Date _____ Class _____

Exploration 1 Worksheet

What's Bugging You?

1. Dr. Mike Roe's patients are suffering from the same symptoms. Describe what Dr. Roe wants you to do so that he can treat his patients.

   _____

   _____

   _____

   _____

2. What is the difference between the two different trays of slides on the front table in Dr. Labcoat's lab?

   _____

   _____

   _____

3. How will you use the microscope and the slides to help Dr. Roe?

   _____

   _____

   _____

   _____

4. What other information will you then need to find out about the six sample organisms?

   _____

   _____

   _____

   _____

5. How could you find such information?

   _____

   _____

   _____

EXPLORATION 1 • WHAT'S BUGGING YOU?   79

Name _____ Date _____ Class _____

Exploration 1 Worksheet, continued

6. Examine the 13 reference slides under the microscope, and record your observations, along with the name of each organism, in the table below.

Name of organism	Observations
1.	
2.	
3.	
4.	
5.	
6.	
7.	
8.	
9.	
10.	
11.	
12.	
13.	

Name _____ Date _____ Class _____

Exploration 1 Worksheet, continued

7. Now examine the six sample slides, and record your observations in the table below. Refer to your observations of the reference slides, and use the third column below to record any similarities you find between the reference slides and each sample slide.

Sample	Observations	Similarities to reference slides
1		
2		
3		
4		
5		
6		

8. In the CD-ROM articles, read about the interactions that each of the six sample organisms have with humans. Record your notes here. Use another piece of paper if necessary.

_____
_____
_____
_____
_____
_____
_____

Record your answers in the fax to Dr. Roe.

Name _____ Date _____ Class _____

Exploration 1
Fax Form

FAX

To: Dr. Mike Roe (FAX 206-555-7272)

From:

Date:

Subject: Identification of Sample Organisms

What are the six organisms taken from Dr. Roe's patients, and what is their classification: mutualistic, commensalistic, or parasitic?

Sample	Name of organism	Mutualistic	Commensalistic	Parasitic
1				
2				
3				
4				
5				
6				

Of the six sample organisms, which is the most likely cause of the patients' fever, chills, and muscle aches?

Sample 1	Sample 2	Sample 3	Sample 4	Sample 5	Sample 6

What is the name of the illness caused by this organism, and how is it transmitted to humans?

_____
_____
_____
_____
_____

What's Bugging You?

The following articles can also be found by clicking the computer in the CD-ROM laboratory for Exploration 1:

- *Communities in Nature*
- *Humans as Hosts*
- *The Spread of Infectious Disease*

Exploration 1
CD-ROM Articles

Communities in Nature

What Makes a Community?

You are probably familiar with the term *community*. A community is a place where people live—a village, town, or city. A **community** in the biological sense is made up of different groups of organisms that live together in a particular place, or **habitat.** Different species may share the same habitat. In a pond habitat, for example, bullfrogs, fish, water bugs, grasses, algae, and other plants and animals form a community. Of course, all communities require other resources for survival, such as air, soil, and water. A community and all of its nonliving parts are called an **ecosystem.**

Finding a Niche

A species' way of life in its habitat is called its **niche.** An organism's niche includes what it eats, what eats it, where it lives, how it uses water and soil, and all the other ways in which the organism interacts with its environment. Although many different species share a habitat, only one species in a habitat can occupy a particular niche. For example, the various species living in and around an oak tree interact with their habitat differently. Squirrels may eat acorns, while birds may eat insects. One species of ant may live inside the tree, while another species burrows in the ground beneath the tree.

Commensalism, Mutualism, and Parasitism

One type of relationship in a community is called **commensalism.** In commensalism, one species benefits from the relationship while the other species is neither harmed nor helped by the interaction. For example, a clownfish can hide from predators within the poisonous tentacles of a sea anemone because the clownfish is immune to the anemone's deadly stings. The clownfish also feeds on bits of food left over from the anemone's meals. The clownfish clearly benefits from the anemone. The sea anemone, however, doesn't receive any noticeable benefit from the clownfish.

Another type of interaction between species in a community is called **mutualism.** In this type of interaction, each species is helped by the mutualistic relationship. One example of mutualism takes place in termites. As you know, termites feed on wood. What you may not know is that termites cannot digest the wood for themselves. To accomplish this, they rely on bacteria that live in their digestive system. In this association, the termite gets nutrition from the digested food, and the bacteria get a safe habitat in which to live.

Parasitism is a third type of relationship among species. In a parasitic relationship, one species benefits and the other is harmed. Parasitism is actually a form of predation. The parasite uses its host as a source of food, a place to live, or a place to raise young. Parasites eventually weaken their host and may even kill it. Ticks, lice, mosquitoes, and tapeworms are common parasites. Many microorganisms are also parasites.

Humans as Hosts

The Human Body

The human body is host to a large number of organisms. Some live on the outside of our body, while others live inside. Many organisms live in warm, moist regions, such as inside the nose and mouth. One species of bacteria lives inside the small intestine. Still other organisms live on flakes of dead skin and hair. Many of the organisms that live on or inside the human body are harmless, and a few are even helpful. Some, however, can be extremely harmful—even deadly.

EXPLORATION 1 • WHAT'S BUGGING YOU?

Chlonorchis sinensis (Human Liver Fluke)

Identification
This organism is 10–22 mm long and 3–5 mm wide. The adult fluke has a flat body with suckers.

Environment
Most infections occur in Asia, although some cases have been reported the United States. Liver flukes are found in the digestive system. Humans may ingest the parasite when they eat raw or undercooked infected fish.

Symptoms of Infection
Flukes feed on cells, blood, and body tissues. If the infestation is severe, massive swelling of the affected tissue may occur. When the liver's function is impaired, **jaundice,** a discoloring of the skin and other body tissues, can occur.

Demodex folliculorum (Follicle Mite)

Identification
This follicle mite lives on the outside of the human body.

Environment
This organism is found on humans throughout the world. It lives in hair follicles, sweat glands, and oil glands on the skin.

Symptoms of Infection
The follicle mite is most commonly found around the face, especially on the nose and eyelids. It normally causes no reaction in humans, although these mites may be linked to mild skin problems such as acne.

Dermatophagoides farinae (House Dust Mite)

Identification
Adult house dust mites have long, oval bodies with four pairs of legs, and their mouthparts look much like a head.

Environment
House dust mites live in vacuum-cleaner bags, mattresses, pillows, and furniture that contains natural plant fibers. These scavengers feed on human skin scales and other organic debris.

Symptoms of Infection
House dust mites don't bite humans, but they can cause allergic reactions. About 90 percent of people who are allergic to house dust are also allergic to dust mites.

Entamoeba gingivalis (also spelled *Entameba gingivalis*)

Identification
This amoeba commonly lives in the gums of the human mouth.

Environment
Entamoeba feasts on bacteria and tiny bits of left-over organic matter, like food, that are small enough for the amoeba to surround and absorb.

Symptoms of Infection
These organisms are so small and so well hidden in human gums that they are hardly noticed. The moist environment of the mouth provides an ideal habitat for these amoebas. *Entamoeba* cannot be brushed or rinsed away, but a normal population of *Entamoeba* causes no reaction in humans.

Enterobius vermicularis (Pinworm)

Identification
Commonly called the pinworm, this organism is an intestinal worm related to the hookworm and is most common among small children, who often practice poor hygiene.

Environment
Eggs of the pinworm can be swallowed when contaminated water or soil is ingested. The eggs pass through the body to the large intestine, where they mature. Sometimes worms that mature at the edge of the anus may reinfect the large intestine or even move to the appendix.

Symptoms of Infection
The movement of the worms causes severe itching, swelling, and discomfort. Scratching the area may break the skin and cause a bacterial infection. Scratching also spreads the pinworms by transferring eggs or larvae from the fingers to clothing or bed linens. A prescription medicine may be taken to expel the worms from the body.

Exploration 1 CD-ROM articles, continued

Giardia lamblia

Identification
This microorganism is a protozoan. It has a tail, called a flagellum, that enables it to move through fluids.

Environment
Giardia lamblia is found in bodies of water worldwide, even in chlorinated drinking water. Humans can contract this protozoa by drinking contaminated water. The feeding form of this protozoa lives in the small intestine.

Symptoms of Infection
An infected person may suffer nausea and vomiting, cramps, diarrhea, and weight loss. The protozoa cause the intestinal lining to swell, interfering with digestion and the absorption of nutrients.

Naegleria fowleri

Identification
This organism is a protozoan that occurs in the human body as an amoeba.

Environment
This free-floating organism lives at the bottom of lakes, streams, hot springs, and swimming pools. It thrives during the warm summer months in regions all over the world.

Symptoms of Infection
Millions of people have been exposed to this organism and have not become ill. However, if the protozoan gets into the human body, a deadly infection can occur. In extremely rare instances, the protozoa are transmitted into the body through membranes inside the nose. *Naegleria* infects the membranes that surround the brain and spinal cord, causing headaches and fever. The organism moves into the brain, rapidly destroying tissue and usually causing death within a week of infection.

Propionibacterium acnes

Identification
Propionibacterium acnes is a kind of bacterium.

Environment
These bacteria live below the surface of the skin on the human body. They are found around hair follicles, sweat glands, and oil-producing glands. The bacteria occur most commonly on the face, neck, and back.

Symptoms of Infection
Propionibacterium acnes is one kind of bacterium that may contribute to acne. However, most people carry this microorganism but do not have acne. Adolescents are most commonly affected, and populations of the bacterium decline as a person ages. *Propionibacterium acnes* prevents the colonization of other harmful organisms. Without them, humans would suffer from the effects of many harmful organisms.

Rickettsia rickettsii

Identification
An unusual kind of bacterium, *Rickettsia rickettsii*, requires a host cell to reproduce and grow.

Environment
These bacteria live in ticks, which can transmit the microorganism by biting animals and people. The bacteria are transmitted through the tick's saliva. The ticks and bacteria are found in the Rocky Mountains, the eastern United States, Mexico, and South America.

Symptoms of Infection
At the site of the tick bite, the bacteria multiply. The bacteria spread through the blood to the rest of the body. Areas commonly affected include the skin, the spleen, and the nervous system. The disease caused by these bacteria is called Rocky Mountain spotted fever, and it was first identified in the 1870s. Symptoms include a high fever and body aches. In rare cases, death results.

Salmonella

Identification
Salmonella is a common kind of bacterium.

Environment
Many animals carry a species of *Salmonella* bacteria. The bacteria exist worldwide.

Symptoms of Infection
When humans ingest *Salmonella*, they may become mildly or seriously ill. Most infections come from contaminated food, particularly chicken, turkey, and dairy products. The bacteria usually invade the cells of the intestinal wall, causing

EXPLORATION 1 • WHAT'S BUGGING YOU?

diarrhea and vomiting. Blood can transport the bacteria to other parts of the body, including the lungs. Proper food storage and basic kitchen sanitation can help prevent *Salmonella*. Eggs and meat should be cooked thoroughly, and counter tops should be disinfected regularly to prevent the growth of *Salmonella*.

Schistosoma mansoni (Blood Fluke)

Identification
This organism develops into a worm 10–20 mm long.

Environment
The larva of the blood fluke is found in the waters of Africa, Madagascar, Brazil, Venezuela, Puerto Rico, and the Dominican Republic. The blood fluke infects the blood vessels and intestines. When the blood fluke lays eggs, they are carried to the intestinal walls and pass out of the body with human waste.

Symptoms of Infection
When the eggs are released from the intestine, they can cause swelling and bleeding. If the liver is severely infected, liver failure can occur. The worms can live in the human body for up to 25 years.

Staphylococcus epidermis

Identification
Staphylococcus epidermis is a kind of bacterium.

Environment
Staphylococcus is commonly found on human skin. Normally, *Staphylococcus* helps prevent other organisms from colonizing on the skin. Occasionally, the bacteria will enter the body through a break in the skin.

Symptoms of Infection
In cases where *Staphylococcus* enters the body through a break in the skin, an infection can occur.

Streptococcus mutans

Identification
Streptococcus mutans is a kind of bacterium that has an outer capsule made from glucose, a simple form of sugar. The bacteria form chains called **plaque**.

Environment
These bacteria occur only on teeth. They thrive in an environment rich in sugars, especially the sugar found in candy and other sweets.

Symptoms of Infection
The formation of plaque contributes to dental caries (cavities). Where plaque-forming bacteria reach beneath the tough surface of the tooth, cavities form.

The Spread of Infectious Disease

Infectious diseases, such as colds, the flu, and chickenpox, are caused by microbial parasites that invade a host and cause it to become weakened. Most of these parasites are microorganisms such as viruses, bacteria, fungi, and protists.

Some diseases are transported through the air. Flu, for example, is caused by the influenza virus, which usually travels in droplets sprayed into the air when an infected person sneezes. When a person nearby breathes in these droplets, this person also becomes infected with the virus.

Some diseases are transported through food and water. One example is dysentery, which causes life-threatening diarrhea and vomiting. It is caused by microorganisms that live in impure water. Food poisoning is another common illness. It can result from bacteria, like *Salmonella,* that grow on foods that are not properly cooked or stored.

Some diseases can be transmitted by animal contact. A person who handles an animal infected with ringworm or mange, for example, may contract the disease. Animal bites can spread other diseases, such as rabies. Insect bites from mosquitoes, ticks, and fleas can transmit a variety of diseases. The deadliest bacterial epidemic in human history, the Black Death, killed tens of millions of people in the fourteenth century. This disease was transmitted by rats that carried fleas infected with the Black Death bacteria.

Some diseases spread through person-to-person contact. Kissing or sharing a drink with someone who has a cold may transmit a disease. Coming in contact with open sores can also pass infections along. The viral disease AIDS, which is carried by HIV, is passed through sexual contact or exposure to infected blood.

Exploration 2
Teacher's Notes

Sea Sick

Key Concepts	Animals in captivity should have surroundings that match those of their natural habitat. Hydrothermal vents provide unique living conditions to which some very unusual creatures have adapted.
Summary	Some unusual animals were delivered to the Marine Exploratorium. Since their arrival, they have not lived up to their reputation of being spectacular creatures. To make them part of a permanent exhibit, Shelley C. Waters needs to know what kind of animal they are and in what living conditions they thrive.
Mission	Help Shelley C. Waters re-create the proper habitat for some very unusual sea creatures.
Solution	The unusual sea creatures are giant tubeworms. A specific environment must be re-created if the tubeworms are to survive in an exhibit. Since the tubeworms' natural habitat is a hydrothermal-vent community in the deep ocean, increasing the pressure and changing the mineral content of the sea water in the tank make the tank environment most like the tubeworms' natural habitat. As a result, the tubeworms appear much healthier.
Background	Some organisms in the deep-oceanic zone have remarkably striking, almost prehistoric appearances. Many deep-ocean fish, with their gaping mouths, fanglike teeth, and large eyes, resemble fossils of fish that lived 60 million years ago. As consumers, the organisms of the deep ocean usually depend on dead and decaying organisms or parts of organisms that fall to the ocean floor from the upper regions of the ocean. When this source of food becomes scarce, however, the predatory organisms become attack animals and make feasts of each other. These killer instincts explain the fishes' vicious mouths and razor-sharp teeth.
	Because no sunlight reaches the depths of the oceanic zone, some fish and squid there have adapted "glow-in-the-dark" capabilities. Some squid can emit a luminous fluid, and some fish have luminous sexual organs that flash on and off. The function of these adaptations is yet undetermined, but some scientists think that they may serve to identify different species, ward off predators, or facilitate mating.

Exploration 2 Teacher's Notes, continued

Teaching Strategies	Encourage students to keep accurate notes about how each change in variable affects the animals. Students may have difficulty associating the microorganisms on the back counter in the lab with the sea creatures. To help students make the connection, instruct them to examine the slides after completing their experimental testing. Ask what these slides have to do with the animals in the tanks. Suggest that they research the CD-ROM articles to learn about the role that these microorganisms play in the world of hydrothermal vents. They will need to understand this information to complete the last question on the fax form. As an extension of this Exploration you may choose to have students learn more about other organisms that live near hydrothermal vents. Have students focus on the different adaptations that these interesting creatures have evolved that enable them to live in such a unique environment.
Bibliography for Teachers	Morrison, Philip. "Gutless (Red-Tipped Tubeworms)." *Scientific American,* 275 (1): July 1996, p. 104. Van Dover, Cindy Lee. "The Depths Illuminated." *Natural History,* 105 (1): January 1996, pp. 72–73.
Bibliography for Students	Dybas, Cheryl Lyn. "The Deep-Sea Floor Rivals Rainforests in Diversity of Life." *Smithsonian,* 26 (10): January 1996, p. 96. Fisher, Charles, and Ian MacDonald. "Life Without Light." *National Geographic,* 190 (4): October 1996, pp. 86–97. Stover, Dawn. "Creatures of the Thermal Vents." *Popular Science,* 247 (5): May 1995, p. 55.
Other Media	*Physical Oceanography* Video or videodisc with teacher's guide SVE (Society for Visual Education) 6677 N. Northwest Highway Chicago, IL 60631 800-829-1900 In addition to the above video, there are many articles about the unusual adaptations of deep-sea organisms on the Internet. Interested students can access relevant articles with keywords such as the following: *hydrothermal vents, deep-sea ecosystems,* and *tubeworms.*

Name _____ Date _____ Class _____

**Exploration 2
Worksheet**

Sea Sick

1. Shelley C. Waters has sent some unusual creatures to the lab. What is wrong with them, and what does she want you to do for her?

2. The front lab table is pretty crowded with equipment! Describe the different parts of the setup.

3. Why is it necessary to use a control in this experiment?

4. What are the settings on the control tank?

EXPLORATION 2 • SEA SICK   89

Name _____ Date _____ Class _____

Exploration 2 Worksheet, continued

5. Conduct your experiments, and record your observations in the table below.

Change in variable	Observations
Add food.	
Add antibacterial water conditioner.	
Increase pressure in tank.	
Change filter.	
Increase temperature.	
Change mineral composition of sea water.	

Name _____ Date _____ Class _____

Exploration 2 Worksheet, continued

6. What other information do you need to give Ms. Waters?

7. Describe how you might learn this information.

8. Examine the slides Dr. Labcoat has set up on the back counter. What you see plays a role in the habitat of the mystery creatures. What is that role? Explain. (Hint: Check out the CD-ROM articles.)

9. Describe the relationship between your experiment results and what you have learned about these organisms and their habitat.

Record your answers in the fax to Ms. Waters.

EXPLORATION 2 • SEA SICK

Name _____ Date _____ Class _____

Exploration 2
Fax Form

FAX

To: Ms. Shelley C. Waters (FAX 619-555-8368)

From:

Date:

Subject: Habitat

Name of organisms:
_____

Where do these animals live?
_____
_____

During your experiments, which of the following changes had a positive effect on the animals?

☐ ADD FOOD.
☐ ADD ANTIBACTERIAL WATER CONDITIONER.
☐ CHANGE MINERAL COMPOSITION OF SEA WATER.
☐ INCREASE PRESSURE IN TANK.
☐ CHANGE FILTER.
☐ INCREASE TEMPERATURE.

What are the nutritional requirements of these animals?
_____
_____
_____
_____

Name _____

HOLT SCIENCE AND TECHNOLOGY INTERACTIVE EXPLORATIONS TEACHER'S GUIDE

> **Sea Sick**
>
> The following articles can also be found by clicking the computer in the CD-ROM laboratory for Exploration 2:
>
> - *Amazing Undersea Finds*
> - *Oasis in the Depths*
> - *Energy for Life*
> - *Some Creative Adaptations*
> - *Controlled Experiments*

Exploration 2
CD-ROM Articles

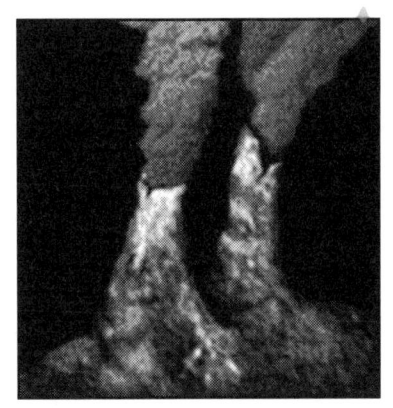

Amazing Undersea Finds

An Accident of Geography

Beneath the ocean are some of the most dramatic landforms on the face of the Earth. Huge mountains, deep canyons, enormous plains, bubbling volcanoes—all are hidden from sight on the ocean floor.

The **mid-ocean ridges,** which are the most prominent features of the ocean floor, are sites of tremendous geologic activity. Crisscrossing the ocean floor like the seams on a baseball, the mid-ocean ridges form the longest mountain range on Earth. They mark the sites where the Earth's crust is splitting apart and new crust is being formed.

The Earth's crust is made up of a number of different sections called **plates.** Forces within the Earth cause these plates to move across the surface in different directions. Moving plates grind against each other, brush past each other, or pull apart. Where the plates pull apart, a crack is opened in the Earth's crust. Molten rock from the interior flows up through the constantly enlarging crack. The molten rock cools to form new crust. This continual process forms the towering mountain ranges of the mid-ocean ridges.

Letting Off a Little Heat

The crust near mid-ocean ridges is very thin and full of cracks called **mid-ocean rifts.** Underwater hot springs, called **hydrothermal vents,** emerge from many of these cracks. These hydrothermal vents provide a warm oasis in the otherwise frigid depths. In addition to being warm, water from hydrothermal vents is also rich in minerals.

Heat escapes from vents in the ocean floor.

The temperature of the water emitted by the vents varies widely. Many vents emit water with temperatures between 5°C and 23°C. Hotter vents release water with temperatures up to 380°C. Despite the high temperature, the water never boils because of the tremendous pressure found at the depths where hydrothermal vents occur.

What Scientists Found Down There

The deep-ocean floor is a dark, barren, almost lifeless place. Because no sunlight can reach the deepest portions of the ocean's floor, plants—the base of every known food chain—cannot exist there. For years, it was assumed that without plants, there could be no other life. But in 1977, scientists made a startling discovery.

Using the deep-sea research vessel *Alvin,* scientists were exploring a section of mid-ocean rift off the west coast of South America, near the Galápagos Islands. Suddenly, at the tremendous depth of about 2500 m, the scientists came upon a site that was teeming with life! The unusual community contained giant tubeworms, huge clams and shrimp, albino crabs, and strange fish, all unlike anything seen elsewhere in the ocean. Later, more of these strange communities were found clustered around the many hydrothermal vents that dot the ocean bottom in the area.

Those scientists discovered a previously unknown ecosystem. This ecosystem was based not on plants—which get their energy from the sun—but on a type of bacteria that manufactures food using chemicals emitted from the hydrothermal vents. This process of making food from chemicals is called **chemosynthesis.**

Oasis in the Depths

A Wealth of Resources

The areas around hydrothermal vents are extremely rich in minerals, making them a potentially valuable resource. As sea water trickles down through cracks in the Earth's crust, it dissolves minerals from the rock. Eventually, the sea water makes contact with magma beneath the crust. This causes the sea water to become superheated and to begin to rise to the surface of the sea floor. As the superheated sea water rises to the surface of the ocean floor, it dissolves additional minerals and chemicals from the rocks. Eventually, the hot, mineral-laden water reaches the sea floor, where it emerges as a hydrothermal vent. The hot water mixes with the cooler ocean water, causing the dissolved minerals to precipitate, or settle out, onto the sea floor.

The minerals associated with hydrothermal vents include metal sulfides, iron oxide, silica-rich compounds, and barium sulfate. Deposits rich in copper, chromium, silver, gold, and platinum are also common.

Life Around a Hydrothermal Vent

Most hydrothermal-vent communities are found at a depth of about 2500 m. The terrain in such a region is rocky. The water is often clouded with minerals. Few organisms live very close to the hottest vents, which can have temperatures as high as 380°C. Most organisms live farther away, in waters with a temperature of about 25°C.

The water pouring out of a hydrothermal vent is rich in a compound called hydrogen sulfide. This compound attracts a type of bacteria that creates a chemical reaction which converts hydrogen sulfide into food. In the hydrothermal-vent ecosystem, these bacteria play the role played by plants in other ecosystems. As a result, the bacteria form the base of the hydrothermal-vent ecosystem.

Interactions in a Vent Community

In the hydrothermal-vent ecosystem, as in all ecosystems, organisms interact with one another. Many organisms have a predator-prey relationship, in which one organism is eaten by another. White crabs, for example, survive by eating tubeworms and mussels. Some vent organisms are filter-feeders. **Filter-feeders** are organisms that feed primarily on dead and decaying organisms that sink down from the upper regions of the ocean. The filter-feeders feed by straining the bits of organic matter from the water.

Other organisms form mutualistic relationships. In these relationships, both of the organisms benefit from the interaction. Tubeworms and bacteria have a mutualistic relationship; the tubeworms provide the bacteria with a habitat inside their bodies. In return, the bacteria supply the tubeworms with food energy.

Energy for Life

The Flow of Energy in an Ecosystem

All living things need energy to survive because every life process requires energy. An ecosystem is, in fact, a system for distributing energy among living things.

In almost all cases, ecosystems harness energy from the sun. Organisms that convert energy from a nonliving source, such as sunlight, into food energy are called **producers.** Producers, which include plants, algae, and phytoplankton, make up an ecosystem's lowest energy level.

Organisms that eat producers are called **consumers. Primary consumers,** called **herbivores,** eat only plants and occupy the second energy level. A herbivore can be as small as a fruit fly or as large as an elephant. Organisms in the next higher energy level are called **secondary consumers.** Secondary consumers are **carnivores**—organisms that eat other animals. Snakes and lions are both carnivores. Organisms, like humans, that eat both plants and animals are called **omnivores.** Both omnivores and carnivores belong to the third energy level of an ecosystem.

Some ecosystems have a fourth energy level that contains a **top carnivore.** The top carnivore generally does not become a source of food for any other organisms except decomposers. **Decomposers** are organisms that break down the remains of once-living things. Worms, bacteria, and fungi are examples of decomposers.

Photosynthesis

Photosynthesis is a process that converts solar energy into chemical energy, which plants can use to carry out life processes. The ingredients for photosynthesis are fairly simple—sunlight, water, carbon dioxide, and chlorophyll.

Chlorophyll is a chemical that absorbs light. It is found in plants and some microorganisms. The energy from sunlight is used to split water molecules into atoms of hydrogen and oxygen. The oxygen is released as a waste product, and the hydrogen is combined with carbon dioxide. Carbon dioxide is made up of carbon atoms and oxygen atoms. The carbon, oxygen, and hydrogen form a simple sugar, which serves as a source of stored energy. This sugar can be converted by an organism into starches, fats, and other compounds necessary for life.

When photosynthesis is complete, the living organism uses the chemical energy to carry out life processes. Extra energy can be stored for future use. Any animal that eats the plant or microorganism also eats the stored energy. In this way, the energy captured from the sun fuels the life processes of all the living organisms in the ecosystem.

Chemosynthesis—Making Food Without Light

Most producers use sunlight as their energy source. However, some recently discovered ecosystems are based on producers that use chemosynthesis to create food energy. **Chemosynthesis** is the process of using chemical energy to make food. In the vent communities, it is the chemical energy from hydrogen sulfide, rather than the sun's energy, that provides the energy to make food.

Bacteria in vent communities combine atoms of hydrogen from hydrogen sulfide in the sea water with carbon and oxygen atoms from carbon dioxide. The carbon, hydrogen, and oxygen atoms form the simple food compounds that provide living organisms with energy. Although hydrogen sulfide is extremely poisonous, giant tubeworms have adapted to the high level of hydrogen sulfide in vent waters. They deliver the chemical to the bacteria that live in their tissues. The bacteria then convert the hydrogen sulfide into food energy, and the giant tubeworm uses the extra energy as food.

Thiothrix sp. are multicellular organisms that live in sulfur-rich environments, such as the rocky surfaces around hydrothermal vents. (Note: The "sp." signifies that this name could refer to any of a number of species.) Large numbers of *Thiomicrospira* sp. are found around hydrothermal vents. They are small bacteria that utilize sulfur compounds. The rod-shaped *Thiobacillus* sp. get their energy from the chemosynthesis of sulfur and sulfur compounds. *Beggiatoa* sp. have filaments that allow the bacteria to move. In most other ways, these bacteria are similar to *Thiothrix* sp. *Vibrio* sp. are a kind of luminescent (light-emitting) bacteria. An enzyme inside the bacteria causes the microorganisms to give off light.

Some Creative Adaptations

In all ecosystems, living things adapt to their environment. Some of the most amazing adaptations—adaptations for protection, food-gathering, and shelter—are found in organisms in the sea.

A clownfish finds food and protective cover in an anemone bed.

The sea anemone has an interesting self-protective adaptation. Its poisonous tentacles can teach a predator to avoid the anemone in the future. Amazingly, the clownfish, a brightly colored fish, lives safely in this otherwise dangerous place. How does it do this? The clownfish is coated with mucus, which protects it from the anemone's stings. In addition, the clownfish can feed on food particles not eaten by the anemone.

Controlled Experiments

Scientific Method

The scientific method is a systematic way of asking questions, performing experiments, gathering data, drawing conclusions, and communicating results. A scientist begins by asking an investigative question, such as "How do birds know when it's time to migrate?" or "Do heavier objects fall faster than lighter ones?" Then he or she collects information or data about the question to form a hypothesis. The **hypothesis** is a possible explanation for an event. A good hypothesis is a statement or explanation that can be tested. The scientist then designs an experiment to test the hypothesis. As the experiment takes place, observations are recorded. By analyzing these observations, the scientist can draw conclusions about the hypothesis and communicate the results.

What if you wanted to find out how much fertilizer is best for growing a potted plant? One way to do this scientifically is to set up a controlled experiment. A **controlled experiment** is one in which only one factor or variable is changed at a time.

In this case, the only variable that changes is the amount of fertilizer each plant receives. Everything else—including the type and size of plant, amount of water, amount and intensity of sunlight, and type of soil—must stay the same. The experimental plants would be exactly like the control plant, except they would be given different amounts of fertilizer. This way, you can easily test the effect of fertilizer on the growth of the plant.

Recording Results

Scientists must carefully record the results of their work. You may decide to record your results in your notepad using a form that looks something like this:

Title of the experiment:

Description of the problem or question:

Research about the problem or question:

Hypothesis:

Variables to be controlled:

Experimental variable:

Description of experiment performed:

Data and observations:

Do these observations support the hypothesis?

Additional questions or possible hypotheses:

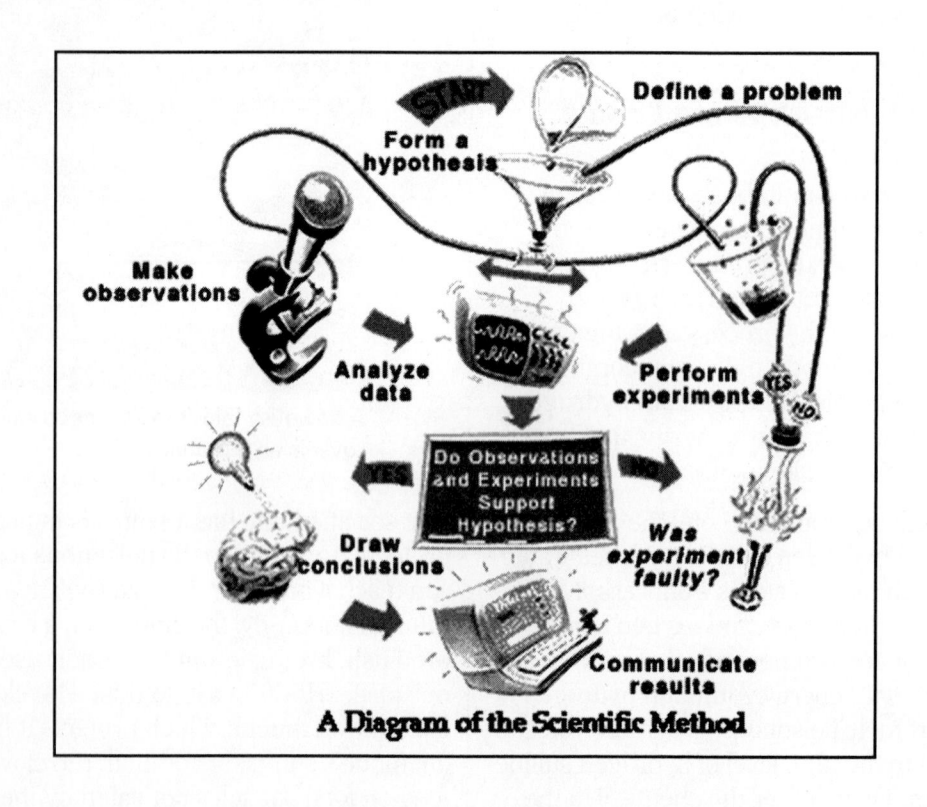

A Diagram of the Scientific Method

Exploration 3
Teacher's Notes

Moose Malady

Key Concepts	Acid rain can damage multiple parts of an ecosystem. Liming, a practice designed to counteract the effects of acid rain, can have counterproductive results.
Summary	A moose population in western Sweden is in trouble. A shortage in blueberries (the moose's normal diet) has led the moose to eat barley and oats from neighboring fields. Unfortunately, the moose have somehow become ill. Hans Oleson has sent samples of items he suspects play a role in the epidemic to Dr. Labcoat's lab for analysis. He wants to know the cause of the moose malady and the reason for the decline in the blueberries.
Mission	Find out why the moose population is ill and why the numbers of blueberries are declining.
Solution	The moose are suffering from a copper deficiency. The moose's internal organs had high levels of molybdenum and low levels of copper because the moose were eating from barley and oat fields that had been limed. Spraying fields with lime raises the pH of the soil. The higher pH allows more molybdenum to mobilize out of the soil, thus immobilizing the copper naturally present. The numbers of blueberries are declining because acid rain has damaged the soil in which blueberry bushes grow.
Background	When fossil fuels are burned, they release oxides of sulfur and nitrogen as byproducts. When the oxides combine with water in the atmosphere, they form sulfuric acid and nitric acid, which fall as precipitation. Power plants are some of the largest contributors of pollutants that cause acid precipitation. When oxygen and nitrogen combine at high temperatures, such as during combustion, nitrogen oxides are formed. Sulfur dioxide, SO_2, is a noxious gas with a characteristic "rotten-egg" smell; it is toxic to both plants and animals. SO_2 is produced mainly by power plants that burn high-sulfur coal (coal that contains iron sulfides) to generate electricity. A large power plant may burn 9,070 metric tons of coal per day. If the coal contains 3 percent sulfur, about 544 metric tons of SO_2 are produced every day. Automobiles produce very little SO_2 but are the source of almost half of the nitrogen oxides released into the air.

Exploration 3 Teacher's Notes, continued

Teaching Strategies	In order for students to pursue the solution to this Exploration, it is imperative that they understand the differences between the two tasks required of them. You may wish to go over the letter from Mr. Oleson with the entire class, making sure that students understand that what is making the moose ill may or may not be the same thing that is causing the disappearance of the blueberries. You may also want to spend some time talking about the indicator solutions and how they will be used in the lab analysis. The solutions on the back counter of Dr. Labcoat's lab are useful tools for explaining the pH level of solutions.
	As an extension of this Exploration, suggest that students learn more about the distribution of acid rain, either in western Europe or in the United States. Encourage students to find out what types of living things are most affected by acid rain. You may wish to explain that the illness of the moose in Sweden is significant because the moose is one of the first large mammals that scientists have identified as being directly affected by acid rain.
Bibliography for Teachers	Hunter, Beatrice Trum. "How Well Do We Absorb Nutrients?" *Consumers' Research Magazine,* 77 (4): April 1994, p. 17.
	MacKenzie, Debora. "Killing Crops with Cleanliness." *New Scientist,* 147 (1996): September 23, 1995, p. 4.
Bibliography for Students	Hildreth, Jim. "Feeding Summer's Soil." *Mother Earth News,* 144: June/July 1994, p. 52.
	Pearce, Fred. "Acid Fallout Hits Europe's Sensitive Spots." *New Scientist,* 147 (1985): July 8, 1995, p. 6.
	Raloff, Janet. "When Nitrate Reigns." *Science News,* 147 (6): February 11, 1995, p. 90.
Other Media	*Acid Rain: The Invisible Threat* Churchill Media SVE (Society for Visual Education) 6677 N. Northwest Highway Chicago, IL 60631 800-829-1900
	Interested students may also find information relevant to the topics in this Exploration by searching the Internet. Suggest that students conduct their search using keywords such as the following: *acid rain, soil, minerals, plants,* and *ecosystems.*

98 *HOLT SCIENCE AND TECHNOLOGY INTERACTIVE EXPLORATIONS TEACHER'S GUIDE*

Name _____ Date _____ Class _____

Exploration 3
Worksheet

Moose Malady

1. Mr. Oleson is very concerned about some moose in western Sweden. What does he want to know?

   _____
   _____
   _____

2. What has Mr. Oleson told you about both the new and the traditional habitat and niche of the moose?

   _____
   _____
   _____
   _____
   _____

3. Mr. Oleson had the internal organs from several of the moose that died tested. What did he learn?

   _____
   _____
   _____

4. Describe the equipment that Dr. Labcoat has set up on the front table.

   _____
   _____
   _____
   _____

5. What is the purpose of an indicator solution?

   _____
   _____
   _____
   _____

EXPLORATION 3 • MOOSE MALADY   99

Name _____ Date _____ Class _____

Exploration 3 Worksheet, continued

6. Conduct your analysis of Mr. Oleson's samples, and record your results in the table below.

Test	Results		
Indicator solution	Sample 1	Sample 2	Sample 3
Aluminum (Al)			
Cadmium (Cd)			
Chromium (Cr)			
Copper (Cu)			
Iron (Fe)			
Manganese (Mn)			
Molybdenum (Mo)			
Nitrogen (N)			
Phosphorus (P)			
Potassium (K)			
Sulfur (S)			
Zinc (Zn)			
pH indicator			
pH level			

7. In terms of your results above, what do *High, Normal,* and *Low* refer to?

   _____

   _____

8. What clues do you get from analyzing the pH levels of the sample solutions? The CD-ROM articles should help you with your analysis.

   _____

   _____

   _____

   _____

   _____

Record your answers in the fax to Mr. Oleson.

Name _____ Date _____ Class _____

Exploration 3
Fax Form

FAX

To: Mr. Hans Oleson (FAX 46-18-555-5374)

From:

Date:

Subject: Mysterious Moose Malady

What is making the moose sick, and what is causing the numbers of blueberries to decline? Explain.

EXPLORATION 3 • MOOSE MALADY 101

Name _____ Date _____ Class _____

Exploration 3 Fax Form, continued

Which two tests provided the strongest evidence to support your answers to the previous questions?

Test selected	Best evidence
Aluminum	
Cadmium	
Chromium	
Copper	
Iron	
Manganese	
Molybdenum	
Nitrogen	
Phosphorus	
Potassium	
Sulfur	
Zinc	
pH level	

What can Mr. Oleson do to help restore the moose's health?

_____
_____
_____
_____
_____
_____
_____
_____
_____

Moose Malady

The following articles can also be found by clicking the computer in the CD-ROM laboratory for Exploration 3:

- *Support From Below*
- *About Acids*
- *Polluted Skies*

Exploration 3 CD-ROM Articles

Support From Below

The Value of Soil

Many of the living organisms on Earth depend directly or indirectly on soil for food. Plants take in valuable nutrients from the soil through their roots. Animals obtain their nutrients by eating the plants or by eating other animals that eat plants. Nutrients are then returned to the soil through the action of certain microbes on dead organisms.

Soil contains air, water, plant and animal matter, and other mineral and organic particles. The composition of these elements in soil changes constantly and can vary widely from one area to another. The type of soil in an area influences how well crops grow there.

The Essential Elements in Soil

All plants require certain basic chemical elements. Without the proper amount of these elements, a plant's health, appearance, or ability to reproduce may suffer. Of the chemical elements plants need, at least 16 are considered essential elements. All essential elements except carbon, hydrogen, and oxygen are obtained from the soil and absorbed by plant roots.

The Big Six

The following elements are considered **macronutrients** because they are required in relatively large quantities for plant growth:

Nitrogen is needed by plants for lush, sturdy growth.

Phosphorus is critical for photosynthesis, plant maturity, healthy roots, and energy transfers within plants.

Potassium helps plants overcome the negative effects of drought, shortages of light, and diseases.

Calcium supports several critical plant functions, including building cell wall tissues, regulating the availability of other nutrients, and building plant proteins.

Magnesium aids in photosynthesis as well as in the plant's use of other nutrients, including nitrogen, phosphorus, and sulfur.

Sulfur helps plants produce proteins and enzymes.

Only a Trace

Elements that plants need in minute quantities only are called **micronutrients** or **trace elements**. The following are critical trace elements:

Zinc is a catalyst that aids in several chemical reactions in plants.

Boron is responsible for at least 16 plant functions, including cell development and division, fruit and flower production, and stem growth.

Manganese aids a plant in fighting diseases. Increasing a soil's acidity will help ensure this nutrient's availability.

Iron aids in the process of photosynthesis and helps plants take in nitrogen.

Copper activates a plant's respiration process and helps it to use iron.

Molybdenum helps provide plants with a simple form of nitrogen they can use readily. When chemically bound to other elements in acidic soils, molybdenum becomes scarce.

Soil Chemistry

Soils can be acidic, alkaline, or neutral. An acidic soil has a relatively low pH, whereas an alkaline soil has a relatively high pH. Some plants will only grow in soil with a certain pH. A change in pH can alter the size, color, and health of a plant. Soils that are very acidic or very alkaline can harm many plants. This is because the acidity or alkalinity can hinder biological and chemical processes, such as a plant's ability to take in nutrients.

Soil conditioners are sometimes used to alter a soil's pH, improve its ability to hold water, and improve soil drainage. Aluminum sulfate, for example, lowers the pH of soil. Calcium carbonate, or lime, raises the pH of soil. Lime also causes chemical reactions in the soil that make nutrients and minerals more soluble. The result is an increased absorption of some nutrients by the plant.

About Acids

Acids

An **acid** is a substance that releases hydrogen ions when dissolved in water. An **ion** is an electrically charged atom or molecule. The more hydrogen ions that are released, the stronger the acid is. Weak acids, such as vinegar and lemon juice, can be handled safely in any concentration. But strong acids, such as nitric acid and sulfuric acid, can corrode metals and stone and dissolve tissues.

The pH Scale

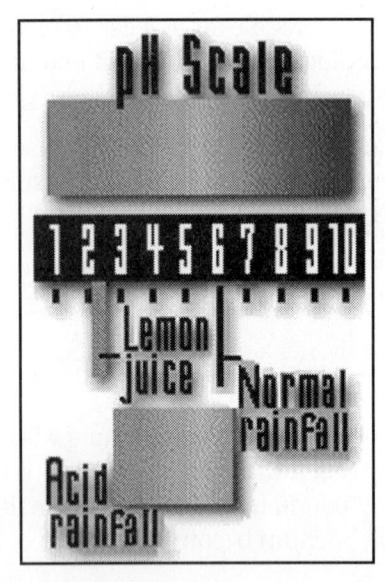

Acidity is measured using the pH scale. The **pH scale** measures the hydronium ion concentration of a solution. Hydronium ions are formed when hydrogen ions, which are released when acids dissolve in water, combine with water molecules.

The pH scale ranges from 0 to 14. Solutions with a pH of less than 7 are said to be acidic. Solutions with a pH of greater than 7 are said to be basic, or alkaline. Solutions having a pH of 7 are said to be neutral. The lower the pH is, the more acidic the solution is. Each whole-number step on the scale represents a tenfold change in acidity. In other words, a solution with a pH of 4 is ten times more acidic than a solution with a pH of 5.

Polluted Skies

Acid Rain

Pollution from power plants, factories, and automobiles can result in acid rain.

Acid rain is caused by pollutants in the air. More specifically, acid rain is produced when water in the atmosphere reacts with pollution from burned fossil fuels, such as coal, oil, and gasoline. When burned, fossil fuels release sulfur and nitrogen oxides, which react with water to form sulfuric acid and nitric acid. Acid rain is essentially a weak solution of sulfuric and nitric acids. Normal rainfall has a pH of about 6, whereas acid rain or snow can have a pH as low as 2. In parts of the northeastern United States, the pH of all precipitation averages about 4.

Troubled by the Wind

Because pollutants can be transported by wind, it is often difficult to determine the precise source of acid rain. Pollutants may travel many kilometers before they react with water in the atmosphere. For example, coal-fired factories in the Midwest are suspected to have caused damaging acid rainfalls in remote parts of Canada.

Exploration 3 CD-ROM articles, continued

Consequences of Acid Rain

Acid rain can damage soil by leaching out important nutrients from the soil. Soil minerals, such as phosphorus, that are not soluble in acidic water can be washed away in runoff. Therefore, under acidic conditions plants may be denied some important nutrients. If the soil in a particular region is already low in nutrients, acid rain can further deplete the soil's nutrients so that the soil can no longer support plant life.

Acid rain can also make certain nutrients and minerals in the soil more available to plants. Normally, minerals such as aluminum, cadmium, and copper are chemically bound in the tiny rock particles that make up soil. However, these minerals are soluble in acidic water, so acid rain chemically releases these minerals from the soil, making them available for absorption by plants. High concentrations of these minerals in plants are toxic and can damage or kill the plants as well as sicken the animals that eat them.

The Ripple Effect

When one part of an ecosystem becomes damaged, the rest of the ecosystem will probably suffer as well. For example, nutrients and minerals leached from the soil by acid rain can poison lakes and rivers. High levels of aluminum ions in lakes can kill fish by damaging their gills. In Sweden, for example, acid rain caused a decrease in the number of fish species and an increase in the growth of algae and fungi.

In northern Europe, increases in the acidity of rainwater and soil have caused a decline in the number of wild blueberry bushes. Moose and other species that rely on the blueberries for food have had to alter their feeding habits. For the first time, moose have started to graze in farmers' fields rather than in the forest. This has created another problem for the moose. Farmers usually take steps to counteract the effects of acid rain on the soil in their fields. The chemicals that the farmers use to lower the acidity of the soil can lead to health problems for the moose.

One Possible Solution

One method commonly used to reduce the effects of acid rain on soil is called liming. **Liming** involves adding calcium and magnesium compounds to soil. This raises the soil's pH, thereby reducing its acidity.

Liming can increase the availability of some nutrients and minerals in soil because some minerals that are not soluble in acidic water, such as molybdenum, are soluble in alkaline water. As a result, alkaline soils can have higher molybdenum concentrations. A change in the mineral content of soil can cause problems for plants and for the animals that eat those plants. For instance, if the levels of molybdenum increase after a soil has been limed, the availability of other nutrients, such as copper, will probably decrease. So animals that eat plants in a field that has been limed may not get enough of the elements they need to stay healthy. Without sufficient copper, for example, an animal can experience serious health problems, such as discolored fur, weight loss, osteoporosis, and, in extreme cases, even death.

The Human Influence

As our use of fossil fuels increases, so do instances of acid rain. In many parts of the world, acid rain has reduced the quality of drinking water and has resulted in decreased production from farms, forests, and fisheries. Acid rain has also been linked to health problems, and it causes excessive damage and wear to sculptures and to the exteriors of buildings.

Preventing Acid Rain

The best way to prevent acid rain from occurring is to reduce the amount of pollutants in the atmosphere. There are many ways to do this, including

- driving less, or taking mass transit or riding bicycles instead of using automobiles,
- heating homes with solar energy or natural gas rather than coal or oil,
- insulating homes to prevent heat loss,
- improving the energy efficiency of automobiles and industry,
- and reducing the output of pollutants from automobiles and industry.

Exploration 4
Teacher's Notes

Force in the Forest

Key Concepts	Some businesses are working to achieve sustainability in the rain forest by sensibly producing and marketing rain-forest products. The acceleration of an object depends on its mass and the initial force (measured in newtons) applied to it.
Summary	Mr. Gustavo Solimões operates a small-scale manufacturing company that specializes in exporting native products from a rain forest in Brazil. Part of the production line contains a 30 m roller track that connects the packing area to the shipping area. Mr. Solimões wants to know how much initial force should be applied to crates of different masses so that the crates make it all the way to the end of the track without falling off and damaging their contents.
Mission	Help Mr. Solimões determine how much force to use to get different crates safely to the end of his production line.
Solution	The appropriate amounts of force necessary to move the crates of 5 kg, 10 kg, 25 kg, and 50 kg, respectively, are as follows: 80 N, 160 N, 400 N, and 800 N. Forces that are too weak for a particular crate will not move the crate all the way to the end of the track, and forces that are too strong for a particular crate will cause the crate to fall off of the track.
Background	*Sustainability* refers to the ability of an ecosystem to meet the needs of the present while maintaining the functions and processes of its natural resources. Thus, a sustainable ecosystem can provide for the needs of future generations. One way some businesses are working toward sustainability is by finding commercial markets for certain rain-forest products, such as Brazil nuts, cashews, coconuts, and beauty products made from copaiba oil. By marketing products like these, companies can create demand for some of the rain forest's natural resources while protecting valuable portions of the rain forest where the raw materials are found. Sustainable businesses grow no more crops than they can sell, and they do not harm the natural resources when harvesting these crops.

Exploration 4 Teacher's Notes, continued

Teaching Strategies

Before beginning this Exploration, you may need to define the word *pneumatic* for your students. Explain to students that a pneumatic pushing device uses compressed air to drive its mechanisms. Examples of pneumatic tools include a dentist's drill, an auto mechanic's wrench, and a jackhammer.

The straightforward nature of this Exploration may increase students' tendencies to quickly conduct their experiments and send in their responses on the computer fax form. To keep students focused on the scientific concepts involved, encourage them to reinforce what they read in the CD-ROM articles by having them conduct force-and-motion demonstrations. Suggest that they use objects such as force meters and wooden blocks or toy cars of different masses to measure the amount of force applied to each object. Students could also explore methods of reducing the effect of friction by conducting the demonstrations on various surfaces or by applying a lubricant such as vegetable oil.

Bibliography for Teachers

Von Baeyer, Hans Christian. "Big G." *Discover,* 17 (3): March 1996, p. 96.

Wade, Bob. "'Hot Wheels' in the Laboratory." *The Physics Teacher,* 34 (3): March 1996, p. 150.

Bibliography for Students

Timney, Mark C. "Ups and Downs of Roller Coaster Physics." *Boys' Life,* 86 (6): June 1996, p. 50.

Pope, Gregory T. "Pipeline Moves Passengers in Ground-Effect Machines," *ScienceNews,* 145 (9): February 26, 1994, p. 143.

Other Media

Cartoon Guide to Physics CD-ROM
Harper Collins Interactive Publishers
10 East 53rd Street
New York, NY 10022
800-424-6234

Total Amazon CD-ROM
CLEARVUE
6465 North Avondale Avenue
Chicago, IL 60631-1996
800-235-2788

Interested students may also find relevant information about force and motion on the Internet. Suggest that students use keywords such as the following to conduct their search: *mechanics, pneumatic tools, force and acceleration,* and *force and newtons.* Students may also wish to access interesting facts about the *rain forest* by exploring the Internet.

EXPLORATION 4 • FORCE IN THE FOREST

Name _____ Date _____ Class _____

Exploration 4
Worksheet

Force in the Forest

1. Mr. Solimões has asked you to help him improve the efficiency of his packing facility. What information, specifically, does he require?

   _____
   _____
   _____
   _____

2. The name of Mr. Solimões's company is Sustainable Rain Forest Products, Inc. What are sustainable rain-forest products, and why might they be important?

   _____
   _____
   _____
   _____
   _____
   _____

3. Describe the lab setup on the front table of Dr. Labcoat's lab.

   _____
   _____
   _____

4. How does this lab setup compare with Mr. Solimões's production line? What is it designed to accomplish?

   _____
   _____
   _____
   _____
   _____
   _____

Name _____ Date _____ Class _____

Exploration 4 Worksheet, continued

5. Conduct all of the possible tests with the lab equipment, and record your results in the table below.

Mass (kg)	Pushing force (N)	Observations
5	80	
5	160	
5	400	
5	800	
10	80	
10	160	
10	400	
10	800	
25	80	
25	160	
25	400	
25	800	
50	80	
50	160	
50	400	
50	800	

6. What other force (besides the pushing force) influences the behavior of the crate on the track? Describe how this force works. Explore the CD-ROM articles if you aren't sure of the answer.

_____
_____
_____
_____
_____

Record your answers in the fax to Mr. Solimões.

EXPLORATION 4 • FORCE IN THE FOREST **109**

Name _____ Date _____ Class _____

Exploration 4
Fax Form

FAX

To: Mr. Gustavo Solimões (FAX 55-11-707-8988)

From:

Date:

Subject: Forces in the Rain Forest

What is a newton?

_____
_____
_____
_____

How much force must be applied to the following masses to successfully send each to the end of the 30 m roller track?

MASS (kg)	FORCE (N)			
5	80	160	400	800
10	80	160	400	800
25	80	160	400	800
50	80	160	400	800

How much force must be applied to a 100 kg mass to send it the full length of a 30 m roller track? How much force must be applied to a 200 kg mass?

_____
_____
_____
_____
_____

> **Force in the Forest**
>
> The following articles can also be found by clicking the computer in the CD-ROM laboratory for Exploration 4:
>
> - According to Newton
> - Newton's Laws of Motion
> - Effects of Acceleration
> - Friction
> - Products of the Rain Forest

Exploration 4
CD-ROM Articles

According to Newton

Force and Motion

Force and motion go hand in hand. A **force** is any push or pull. When a net force is applied to an object, the object will experience a change in its **motion.** Motion is a change in position relative to a reference point. When you throw a ball, you exert a force on it that makes it move. When the ball is moving through the air, its motion can be described by its **acceleration,** or change in its speed or direction. Once the ball leaves your hand, you are no longer exerting a force on it, but it is being acted upon by other forces. Gravitational force pulls it toward the Earth, and friction, due to air resistance, slows it down. These forces will return the ball to a state of rest. Once the ball lands on the ground, it will remain at rest until another force is applied to it.

Sir Isaac Newton and Newtons

Sir Isaac Newton was an English scientist who pondered the relationship between force and motion. His questions, hypotheses, and conclusions led to a system of laws, referred to as Newton's Laws of Motion, that describes how and why objects move. In his honor, the basic unit of force is called the newton.

A **newton** (expressed as N) is the international metric unit used to measure force. 1 N is the amount of force necessary to accelerate a mass of 1 kg by a velocity of 1 m per second for every second that the mass is in motion (or 1 m/s^2). For example, it takes 1 N to make a 1 kg ball accelerate from 8 m/s to 9 m/s in 1 second.

Newton's Laws of Motion

Newton's First Law of Motion

Newton's first law of motion has to do with inertia. **Inertia** is the tendency of an object to resist any change in its velocity. This means that an object at rest, like your desk, will not begin to move by itself. You must push or pull it in order to move it. Similarly, a moving object, like a ball flying through the air, does not stop or turn by itself. Forces cause it to turn and bring it to a state of rest.

Anything that has mass has inertia. In fact, inertia depends on mass. The more massive the object, the more inertia it has. A locomotive, for example, has a large amount of inertia. It takes powerful engines to move a train and heavy-duty brakes to stop it.

Newton's Second Law of Motion

Newton's second law of motion describes the relationship between the force applied to an object and the acceleration produced by that force. Consider what happens when you throw a golf ball and when you throw a basketball. How do you think the force needed to accelerate the golf ball compares with the force needed to accelerate the basketball by the same amount? Since the basketball is more massive than the golf ball, the basketball requires more force than would be applied to the golf ball to match the acceleration of the golf ball.

EXPLORATION 4 • FORCE IN THE FOREST **111**

The following equation mathematically expresses Newton's second law of motion:

Force = mass × acceleration

If you wanted to know how much force is required to accelerate a 10 kg object by 5 m/s², you could use this equation:

Force = 10 kg × 5 m/s²
Force = 50 kg · m/s²
Force = 50 N

How much force is necessary to accelerate the same 10 kg object by 6 m/s²?

Force = 10 kg × 6 m/s²
Force = 60 kg · m/s²
Force = 60 N

As you can see, the larger the force applied to an object, the greater its acceleration will be.

Newton's Third Law of Motion

Newton's third law of motion describes how all forces act in pairs. Another way of thinking about forces in pairs is to consider forces as actions and reactions. Every action—or force—produces an equal reaction in the opposite direction. For example, when a bat hits a baseball, the bat exerts a force on the ball that's directed out toward the field, and the ball exerts an equal force on the bat that's directed inward toward the catcher. If both objects are exerting equal forces on each other, then why does the ball soar away from the bat after it is struck? The answer lies in the fact that the baseball has unbalanced forces acting on it. One of these forces acts on the baseball, causing it to accelerate toward the playing field. The other equal and opposite force acts on the bat, slowing the bat, while having no effect on the ball. Why doesn't the bat fly away in the opposite direction from the ball? The batter is exerting yet another force on the bat, which keeps it from flying away!

Effects of Acceleration

Wow, That's Fast!

What do a car turning a corner, a runner racing out of the starting block, and a bicyclist slowing to a stop have in common? They are all examples of objects that have changed velocity and have therefore accelerated. Objects accelerate when they speed up, slow down, or change direction.

Consider this situation. From a standstill, a cheetah (eyeing a zebra) reaches a velocity of 24 m/s in just 6 seconds. What is the cheetah's average acceleration?

To find out, you would perform the following calculation:

Average acceleration (a) = (final velocity − initial velocity) ÷ time

a = (24 m/s − 0 m/s) ÷ 6 s
a = 24 m/s ÷ 6 s
a = 4 m/s²

So the cheetah can increase its velocity by an average of 4 m/s during each second that it chases the zebra (until it reaches its maximum velocity). A zebra can only accelerate by an average of 2 m/s². Obviously, a zebra needs a big head start to outrun a cheetah!

Too Much Acceleration

Imagine you are riding in a car or on the bus and the driver suddenly speeds up. What happens? Because of your inertia, your body wants to remain at rest, so it feels like your body is pushing against the back of your seat. What happens if the driver slams on the brakes? Again, your inertia makes your body want to remain in motion, so you continue to move forward until another force stops you.

Now imagine that the driver turns a corner too sharply. Your body will lean to one side or the other because it wants to continue moving in a straight line. Changing directions at an extremely fast velocity can be dangerous. Cars can roll over, trains can derail, and bicycles can skid and topple over.

Friction

Frictional Force

There is a force at work between the soles of your shoes and the ground. When you walk, your feet exert a force on the ground, and the ground pushes back on your feet. The contact between your shoes and the ground creates friction. **Frictional force** is the force that opposes motion between two surfaces that are touching. In fact, any time two objects rub against each other frictional force opposes motion and slows objects down or prevents them from moving easily.

Different Amounts of Friction

Certain surfaces can create more friction than others. Imagine sliding your hand along a piece of sandpaper. Now imagine sliding your hand across a surface of polished wood or a metal pipe. Generally, there is less friction between smooth surfaces than there is between rough surfaces.

Lubricants help bicycle parts move more smoothly.

Friction between surfaces can often be reduced by adding a lubricant, such as oil or grease. For example, the oil in a car engine reduces friction between the moving parts of the engine. Without oil, friction between the moving parts of the engine would generate so much heat that the engine could be quickly ruined.

Products of the Rain Forest

What Makes a Tropical Rain Forest?

Tropical rain forests are ecosystems that occur along equatorial regions in Africa, Asia, and Central and South America. These forests are characterized by abundant rainfall, year-round warmth, and continuously high humidity. Rain forests are homes to fantastically diverse communities of plants and animals; scientists estimate that up to 90 percent of the world's species live in tropical rain forests.

The State of the Rain Forests

In many countries, rain forests are being destroyed at an alarming rate. Logging is a lucrative business in many of these areas because of the abundant high-quality wood. Some areas of rain forests are cleared to make room for farms and ranches. Unfortunately, once the land is cleared, the farmers have difficulty sustaining crops in the poor soil.

Moving Toward Sustainability

It is possible to harvest an ecosystem's resources without harming that ecosystem in the process. This is called sustainable use, or **sustainability.** People who rely on the forest for economic reasons can harvest renewable rain-forest products such as fruits, nuts, herbs, plant oils, and latex rubber. Selling these products can be more profitable in the long run than selling wood from rain-forest trees or clearing trees to grow crops.

Extracts from the rosy periwinkle found in Madagascar's tropical forests have been used in the treatment of cancer.

In many ways, the world relies on the diverse bounty of the rain forest. Plant products from the rain forest have been used to create many products, from lotions to medicines. More than 3000 rain-forest products have been identified as possible sources of cancer-fighting chemicals. Rain-forest plants may also be cultivated to create new food sources for humans.

Exploration 5
Teacher's Notes

Extreme Skiing

Key Concepts	When selecting a material for a structure, the effects of compressive, tensile, and shear forces on the material must be considered. A prosthetic limb for a ski racer must be able to withstand the stresses of extreme situations.
Summary	Ludwig Guttman, the director of a leading manufacturer of custom-designed machines and appliances, seeks to develop a special lower-leg replacement for a ski racer who plans to compete in the Paralympic Games. The limb needs to be lightweight but able to withstand the heavy stresses caused by the strenuous physical activity of downhill skiing. Dr. Labcoat needs help determining which of a number of materials would be the best to use in the manufacture of this prosthetic limb.
Mission	Select the best material for a prosthetic limb by testing the shear, compressive, and tensile strengths of various materials.
Solution	Constructing the structural core of the lower-leg replacement out of carbon fiber produces a lightweight, corrosion-resistant, and extremely strong artificial limb that can withstand the stresses of downhill skiing. Titanium alloy and aluminum alloy are also practical choices, but they are not quite as strong or lightweight as the carbon fiber. Oak and magnesium alloy are not strong enough to resist the stresses of downhill skiing, and steel and tungsten alloy are too heavy to be practical materials for an artificial leg.
Background	A *prosthesis* is a synthetic replacement for a missing body part. Many prostheses are surgically attached to bone or joint material, while others operate by external batteries. These batteries are powered by electrical impulses triggered by the contraction of muscles. An artificial limb attached to remaining tissues can often restore movement and allow a person to perform essential, daily tasks as well as a variety of strenuous physical activities. With new technology, scientists and surgeons are getting closer and closer to developing prostheses that approximate the movements and functions of natural limbs. Some surgeons employ robots to assess the size, shape, and materials of the bone to which a prosthesis will be attached.

Exploration 5 Teacher's Notes, continued

Teaching Strategies

Successful completion of this Exploration depends on the students' ability to discern the advantages and disadvantages of using certain materials for a prosthetic limb. Emphasize the importance of evaluating both the strength and the density of each material. For example, the tungsten alloy and the titanium alloy can withstand similar amounts of force. However, tungsten is more than four times as dense as titanium and would therefore be a much less practical choice for a prosthesis. If students are uncertain about how to calculate density, refer them to the CD-ROM articles. Remind them that when using water to find volume by the displacement method, $1 \text{ cm}^3 = 1 \text{ mL}$. By researching the articles, students can also find clues that will lead them to recommend a corrosion-resistant material.

As an extension of this Exploration, you may want to have students learn more about the manufacture of prostheses. Students could present their research in the form of a model diagram that shows how a prosthesis operates. You could also suggest that students find out about other common structural uses for the materials introduced in this Exploration.

Bibliography for Teachers

Hawkins, Dana. "Inventing the Future." *U.S. News & World Report,* 118 (11): March 20, 1995, p. 100.

Nichols, Mark. "High-Tech Limbs." *Maclean's,* 108 (11): March 13, 1995, p. 52.

Reinstein, Leon. "Rehabilitation in the 21st Century." *World Health,* 47 (5): September/October 1994, p. 24.

Bibliography for Students

Fillon, Mike. "Feelings." *Popular Mechanics,* 172 (12): December 1995, p. 78.

Hogan, Dan. "High-Tech Limbs Help Disabled Athletes Set New World's Records." *Current Science,* 82 (7): November 29, 1996, pp. 10–11.

Vogel, Stephen. "Better Bent Than Broken." *Discover,* 16 (5): May 1995, pp. 62–67.

Other Media

Newton's Apple Life Sciences—Hip Replacement
Videodisc
GPN
P. O. Box 80669
Lincoln, NE 68501-0669
800-228-4630

In addition to the above videodisc, students may find relevant information about prostheses and structural forces by exploring the Internet. Suggest that students conduct their search with keywords such as the following: *prosthesis, materials and structures, structural engineering, occupational therapy,* and *orthotics.*

Name _____ Date _____ Class _____

Exploration 5 Worksheet

Extreme Skiing

1. Mr. Ludwig Guttman needs your help. What information has he requested?

   _____
   _____
   _____
   _____

2. Examine the devices on the front table and the back counter in Dr. Labcoat's lab. What are they designed to do?

   _____
   _____
   _____

3. What are the differences between compressive, tensile, and shear forces? If you aren't sure, explore the CD-ROM articles.

   _____
   _____
   _____

4. Conduct all the necessary tests with the device on the front table in the lab, and record your results in the table below.

   Amount of Force Each Material Can Withstand (10^6 N/m^2)

Material	Compressive	Tensile	Shear
Aluminum alloy			
Carbon fiber			
Magnesium alloy			
Oak			
Steel			
Titanium alloy			
Tungsten alloy			

HOLT SCIENCE AND TECHNOLOGY INTERACTIVE EXPLORATIONS TEACHER'S GUIDE

Name _____ Date _____ Class _____

Exploration 5 Worksheet, continued

5. Why do you think the figures for shear strength are usually lower than those for compressive or tensile strength?

   _____

   _____

   _____

   _____

6. What other information do you need to know before recommending a material to Mr. Guttman?

   _____

   _____

   _____

7. How can you determine this information?

   _____

   _____

8. Complete the following table:

Material	Mass (g)	Volume (mL)	Density (g/mL)
Aluminum alloy			
Carbon fiber			
Magnesium alloy			
Oak			
Steel			
Titanium alloy			
Tungsten alloy			

9. What do you notice about the volume of each material? Why might this be the case?

   _____

   _____

   _____

   _____

Record your answers in the fax to Mr. Guttman.

EXPLORATION 5 • EXTREME SKIING   117

Name _____ Date _____ Class _____

Exploration 5
Fax Form

FAX

To: Mr. Ludwig Guttman (FAX 404-555-3296)

From:

Date:

Subject: Material Recommendation

Please indicate your material selection here: _____

Why did you recommend this material?

_____

_____

_____

_____

_____

Please complete the following chart:

MATERIAL	Compressive 10^6 (N/m^2)	Tensile 10^6 (N/m^2)	Shear 10^6 (N/m^2)	Density g/mL
Aluminum alloy				
Carbon fiber				
Magnesium alloy				
Oak				
Steel				
Titanium alloy				
Tungsten alloy				

> **Exploration 5**
> **CD-ROM Articles**

> ### Extreme Skiing
> The following articles can also be found by clicking the computer in the CD-ROM laboratory for Exploration 5:
> - *The Hard-Working Leg*
> - *The Strength of Materials*
> - *Selecting the Right Material for the Job*
> - *Paralympics—Challenges and Champions*

The Hard-Working Leg

Skeletal Structure

Bones are what give the human leg its structural strength. The knee allows the leg to bend in a back-and-forth direction much like a door on its hinges. The bones of the leg are held together by connective tissues, called **ligaments,** that act like straps or rubber bands to stabilize the bones as the leg moves. By connecting the upper and lower leg bones, ligaments help keep the bones from moving from side to side.

Muscle Power

Muscles are connected to bones by **tendons,** another form of connective tissue. The leg muscles work in pairs. In order for the leg to move, one muscle must contract, and the other muscle must relax. **Flexor muscles** cause the leg to bend at the knee, and **extensor muscles** cause the leg to straighten. So one set of muscles pulls the bone in one direction, and the other set pulls the bone back.

Stressing the Leg

Standing, walking, running, and even bending all place stress on the leg. When standing still, each leg bears half the weight of your body. Running puts even more force on the leg.

The leg's structural strength comes from its bones, ligaments, and tendons. Bones are extremely strong—some can withstand a pressure of 2200 Newtons per square cm. In addition, bones are somewhat flexible and lightweight. Ligaments and tendons are tough and elastic.

The Human Lever

A **lever** is a type of simple machine that consists of a bar that pivots about a fixed point called a **fulcrum.** The part of the lever between the fulcrum and the load is called the **load arm,** and the part of the lever between the fulcrum and the point where force is applied is called the **effort arm.** A long effort arm requires less force to move a large load than does a small effort arm.

The human leg is actually a lever. What part of the leg do you think is the fulcrum? the load arm? the effort arm? Read on to find out.

There are three classes of levers, based on the location of the fulcrum, load arm, and effort arm.

In a **first-class lever,** the fulcrum is between the load arm and the effort arm. Examples include a seesaw and a balance scale.

In a **second-class lever,** the effort and load arms are also on the same side of the fulcrum. The load arm is between the fulcrum and the effort arm. An example is a wheelbarrow—the wheel is the fulcrum, the handles are the effort arm, and the load arm is the bucket.

In a **third-class lever,** the effort arm and the load arm are on the same side of the fulcrum, but the effort arm is between the fulcrum and the load arm. Examples include a catapult and the human leg. In the lower leg, the knee is the fulcrum, the lower leg above the foot is the load arm, and the lower leg below the knee (above the load arm) is the effort arm. The leg muscles apply the effort in the human leg.

Artificial Legs

An artificial limb, called a **prosthesis,** is used to replace a missing limb. Artificial legs are perhaps the most common prostheses. In the past, an artificial leg may have consisted of a simple wooden peg. Today, **prosthetics,** the branch of medicine dealing with artificial limbs, has made great strides toward developing stronger, more comfortable, and more functional artificial limbs.

EXPLORATION 5 • EXTREME SKIING

Exploration 5 CD-ROM articles, continued

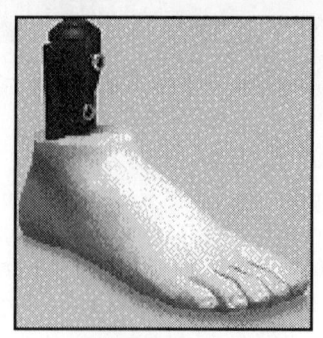

A prosthetic foot

Most artificial limbs are custom-made to fit the individual user. Each part of the prosthesis is specially designed to resist the forces of normal use. Specialists create models on computers to determine the best design for the artificial limb. For example, the toe and the heel respond differently as a person walks. The heel receives the initial force of a person's weight, and the toe pushes off of the ground as the person continues walking. Hence the heel and the toe of an artificial foot must be designed to accommodate these differences. Artificial limbs can be made out of lightweight plastics, fiberglass, wood, metal, and other materials.

The Strength of Materials

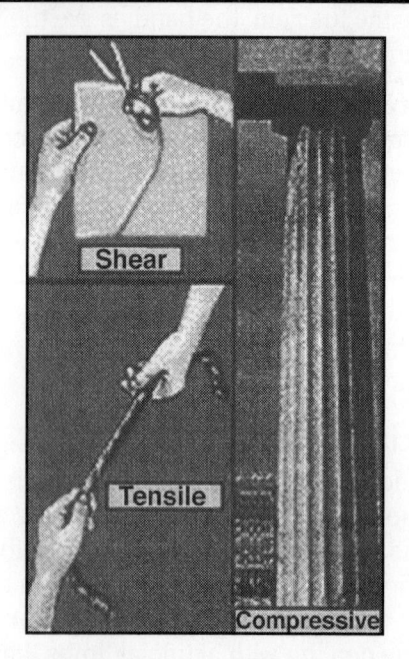

Shear
Tensile
Compressive

Effects of Forces on Materials

Different forces affect materials in different ways. **Compressive forces** are the pushing forces that cause a material to squeeze together. **Tensile forces** are the pulling forces that cause a material to stretch. **Shear forces** are the sideways or twisting forces. Every material will eventually fail if enough force is applied to it. However, some materials handle certain forces better than others. For example, a steel wire withstands tensile forces better than compressive forces.

Properties of Materials

Strength—a material's ability to withstand force without breaking. Strength also affects how much a material deforms under stress.

Density—a material's mass per unit volume, usually measured in grams per cubic centimeter. Materials with a density less than 1 g/cm^3 will float on water.

Elasticity—a material's ability to return to its original shape after being deformed. A rubber band has high elasticity, while a piece of cement has relatively low elasticity.

Flexibility—a material's ability to bend. Rubber has high flexibility because it can be bent easily without breaking. Materials that are not very flexible, like chalk, are described as rigid or brittle.

Ductility—a material's ability to be drawn into wire. Precious metals like gold and silver are highly ductile.

Characteristics of Materials

Aluminum Alloy
Aluminum is a soft, lightweight metal that is highly ductile and resistant to corrosion and wear. Aluminum alloys consist of at least 75 percent aluminum plus one or more elements, such as copper, silicon, and zinc. These alloys are often used for beverage cans, household utensils, and aircraft and automobile parts.

Carbon Fiber
Carbon fiber is a tough, thin fiber that is made by exposing various organic materials to high temperatures and then combining these materials with various synthetic resins. This strong, lightweight, and corrosion-resistant fiber is often used in the construction of spacecraft.

Magnesium Alloy
Magnesium is the lightest of all structural materials. It does not resist corrosion very well because it is a highly reactive chemical. Magnesium alloys are used for lightweight structural parts and batteries.

Oak

Oak is extremely strong for an organic material. It has a distinct grain and is mostly used for furniture. Because oak is subject to natural damage, like wood rot, it is not very corrosion resistant.

Steel

Steel is a metal alloy that is composed of iron and a small percentage of carbon. Sometimes steel also includes small percentages of other metals, such as nickel, chromium, and manganese. Very strong and durable, steel is often used in heavy machinery. Many forms of steel, especially those that contain molybdenum, are very corrosion resistant.

Titanium Alloy

Titanium is a silvery solid used in alloys, powder metallurgy, and the production of pure hydrogen. Titanium alloys resist corrosion and withstand high temperatures, making them especially useful for airplanes and spacecraft.

Tungsten Alloy

Tungsten is a very hard, brittle, highly conductive element that is commonly used for filaments in electric light bulbs and in heating elements. Tungsten alloys, such as tungsten steel, are heavy and corrosion resistant.

Selecting the Right Material for the Job

Testing Materials

Before a material is selected to perform a job, it is usually stress-tested until it fails. A test for tensile strength measures how much force it takes to stretch a material until it breaks or is no longer useful. A machine attaches a load to a material, such as a steel cable. The load is then gradually increased, stretching the cable more and more. At some point, the cable will no longer return to its original shape. This means that the material's elasticity has been overcome, and its tensile strength can be measured.

A test for compressive strength involves "squeezing" the material. The material is placed in a device that puts increasing vertical force on it. The material will become shorter and shorter under greater and greater compressive forces until eventually it fails or is crushed. A machine records the force being applied at the moment of failure. This figure is the material's compressive strength.

To test a material's resistance to shear force, a "twisting" force is applied to the material. Increasing the force eventually causes the material to separate into layers or tear apart. The material's shear strength is measured at the moment of failure. Most structural failures occur when materials break due to excessive shear force.

Making Choices

When selecting component materials for structures or devices, engineers and manufacturers must consider the properties of those materials, particularly how they respond to forces. For example, concrete has low tensile strength and flexibility, but has high compressive strength. It would not be a good substance to use as an elevator cable, but it is an excellent substance to use for the foundation of a building or as a bridge support.

Engineers and architects select building materials best suited for their structures.

Strength is not the only factor that must be considered when choosing a material to do a job. Some materials are extremely strong, but extremely expensive. Other materials may be too dense for a particular task. For example, steel is stronger than aluminum, but an airplane constructed of steel would be too heavy, or too dense, to fly. Elasticity and ductility are other properties that should be considered when choosing a material to perform a job.

Paralympics—Challenges and Champions

Just like the Olympic games, the Paralympics occur every four years in cities across the globe. These games host highly-trained athletes from around the world who compete fiercely to win in their respective events. To be a paralympian, you must have a physical disability. Hundreds of physically challenged competitors gather to compete in events such as swimming, wheelchair basketball, archery, fencing, powerlifting, tennis, volleyball, and soccer.

Many paralympians achieve feats comparable to those of non-disabled competitors. For example, single-leg amputee Arnie Boldt can complete a high jump of 2.04 meters. This height is within half a meter of the height jumped by Charles Austin, the gold medalist in the 1996 Olympic Games.

Rock On!

Key Concepts	The rock cycle is the continuous process of change in which new rocks are formed and old rocks are changed. Rocks can be classified as igneous, sedimentary, or metamorphic.
Summary	Claudia Stone is the superintendent of a new state park. She has sent some rock samples to Dr. Labcoat's lab that she would like identified as igneous, sedimentary, or metamorphic. She would also like to find out about the relationship between igneous, sedimentary, and metamorphic rocks. Ms. Stone plans to use this information to supervise the construction of an information kiosk for park visitors that highlights the unique geological history of the park.
Mission	Provide information that will be used to create a kiosk that highlights the interesting geological features of a new state park.
Solution	Specimens 1, 2, 3, and 8 are igneous rocks; specimens 4, 7, and 9 are sedimentary rocks; and specimens 5, 6, and 10 are metamorphic rocks. Igneous, sedimentary, and metamorphic rocks are related by their participation in the rock cycle, a continuous cycle in which, under specific conditions, any rock type can be converted into any other rock type.
Background	Much about the natural history of an area can be learned by studying its rock formations. For example, the rocks of Glacier National Park in Montana show alternating periods of calm sedimentation and violent upheaval. Most of Glacier National Park's rocks are about a billion years old. Ancient red and gray-green mudstones and siltstones reveal amazing details about the environment in which they formed. Some have ripple marks made by wave motion, while others show mud cracks and impressions made by raindrops. There are also rocks that contain stromatolites (fossilized beds of algae), which indicate that there was a time when this part of the continent was a shallow tidal flat at the edge of a sea. The mud cracks indicate that the sea must have dried out on occasion. There are also thick limestone deposits in the area, indicating periods when the area was a deep-water marine environment. Even evidence of flowing lava is found in the park. Iron-rich magma once forced its way into cracks in the limestone, leaving dark vertical dikes and horizontal sills.

Exploration 6 Teacher's Notes, continued

Teaching Strategies

In order for students to accurately classify the rock specimens in the lab, they must look very carefully at each rock. As students examine each specimen, emphasize the importance of texture, grain size, and composition in determining rock type. Make sure that students record their observations about these characteristics for each rock. You may want to work through one of the more difficult examples, like gabbro, with the class. Encourage students to review the material in the CD-ROM articles and the rock-cycle diagram in the lab before answering the questions on the fax form. You may need to steer students back in the right direction if they become distracted by the UV-light and fluorescent rocks on the back table.

As an extension of this Exploration, encourage students to form groups and choose a state or a national park to study. Have the groups find out about the types of rocks in the area as well as other geological features. Then have students make general conclusions about the geological history of the area based on their findings. Students could report their conclusions in a visual display, such as a miniature kiosk.

Bibliography for Teachers

Olroyd, David. *Thinking About the Earth: A History of Ideas in Geology.* Cambridge: Harvard University Press, 1996.

Earth at Hand—A Collection of Articles from NSTA's Journals. Arlington, VA: National Science Teachers Association, 1993.

Bibliography for Students

Busbey, Arthur B., III, Robert R. Coenraads, David Roots, and Paul Willis. *Rocks and Fossils: A Nature Company Guide.* San Francisco, CA: US Weldon Owen Inc., 1996.

Krause, Barry. "Rock Stars!" *Boys' Life,* 84 (11): November 1994, p. 42.

Other Media

Geology
CD-ROM
National Geographic Society
Educational Services
P.O. Box 98018
Washington, DC 20090-8018
800-368-2728

In addition to the above CD-ROM, students may also find relevant information by exploring the Internet. Suggest that students search the Internet with keywords such as *Earth science; sedimentary, igneous, and metamorphic rocks;* and *rock cycle.* Students may find interesting information about *national and state parks* on the Internet as well.

Name _____ Date _____ Class _____

Exploration 6
Worksheet

Rock On!

1. Claudia Stone is supervising the development of an information kiosk at a new state park. What does she need to know to complete the project?

   _____
   _____
   _____
   _____
   _____

2. What kinds of things are available in the lab to help you provide Ms. Stone with the information she needs?

   _____
   _____
   _____
   _____
   _____

3. Briefly describe the differences between igneous, sedimentary, and metamorphic rocks. Use the CD-ROM articles to help you.

   _____
   _____
   _____
   _____
   _____
   _____
   _____
   _____
   _____

EXPLORATION 6 • ROCK ON!

Name _____ Date _____ Class _____

Exploration 6 Worksheet, continued

4. As you examine each of the rock specimens, record your notes for each rock in the table below. Be sure to show the steps you took through the Identification Key. For example, conglomerate is (1a) grainy and made of more than one material → (2b) particles held together by natural cement → sedimentary.

Rock Classification Table

Rock	Name	Steps from identification key	Classification
1			
2			
3			
4			
5			

126 HOLT SCIENCE AND TECHNOLOGY INTERACTIVE EXPLORATIONS TEACHER'S GUIDE

Name _____ Date _____ Class _____

Exploration 6 Worksheet, continued

Rock Classification Table, continued

Rock	Name	Steps from identification key	Classification
6			
7			
8			
9			
10			

5. What other information do you need to give Ms. Stone?

_____

_____

_____

EXPLORATION 6 • ROCK ON! 127

Name _____ Date _____ Class _____

Exploration 6 Worksheet, continued

6. How could you find this information?

   _____
   _____
   _____

7. Which rock specimens were the easiest to classify? the most difficult to classify? Give reasons for your choices.

   _____
   _____
   _____
   _____
   _____
   _____
   _____

8. Examine the samples on the back counter of the lab. What is unusual about them, and what is the explanation?

   _____
   _____
   _____
   _____

9. What other information could you recommend that Ms. Stone provide in the information kiosk?

   _____
   _____
   _____
   _____
   _____
   _____
   _____
   _____

Record your answers in the fax to Ms. Stone.

Name _____ Date _____ Class _____

Exploration 6
Fax Form

FAX

To: Ms. Claudia Stone (FAX 719-555-0612)

From:

Date:

Subject: Rock Classification

Please classify specimens 1 through 10.

Specimen number	Specimen name	Classification		
		Sedimentary	Igneous	Metamorphic
1	obsidian			
2	gabbro			
3	granite			
4	limestone			
5	marble			
6	quartzite			
7	sandstone			
8	basalt			
9	shale			
10	slate			

Describe the relationship that exists among sedimentary, igneous, and metamorphic rocks.

_____
_____
_____
_____
_____
_____

DISC 2

EXPLORATION 6 • ROCK ON! 129

> **Rock On!**
>
> The following articles can also be found by clicking the computer in the CD-ROM laboratory for Exploration 6:
> - The Rock Cycle
> - Rock On!
> - A Geologic Legacy

Exploration 6
CD-ROM Articles

The Rock Cycle

What Are Rocks?
A **rock** is a combination of two or more minerals. A **mineral** is a naturally occurring solid found on Earth. A mineral has a definite chemical composition and definite physical characteristics. The types of minerals in a rock determine a variety of characteristics, such as the rock's color, resistance to weathering, and melting point.

Metamorphic

Sedimentary

Igneous

Molten rock, called **magma,** forms deep within the Earth, where temperatures and pressure are high enough to melt solid rock. Magma can erupt onto the surface of the Earth through a volcano. The molten rock that spews into the air and onto the Earth's surface is called **lava.**

Rocks are classified by the way they form. To identify and classify rocks, you must be able to recognize some basic characteristics of the three rock types.

- **Igneous rocks** form from the cooling and solidification of magma or lava.
- **Sedimentary rocks** form when particles of rocks and minerals become cemented together. These particles are deposited by water, wind, ice, and chemical reactions.
- **Metamorphic rocks** are rocks that have undergone change as a result of intense heat and pressure.

A Journey Through the Rock Cycle
Did you know that an average rock may have changed in appearance several times during its existence? The **rock cycle** is the continuous process of change by which new rocks form from old rocks. Any rock type—igneous, sedimentary, or metamorphic—may change into any other type under the right conditions. Consider the following imaginary journey:

A Cycle Begins . . .
Imagine that you are standing on the bank of a river. You pick up a rock and toss it into the water. What do you think will happen to it? Follow that rock as it moves through the rock cycle, a journey that takes millions of years.

Stop 1
As the rock is transported by the river, it is worn away over time until it breaks into tiny pieces. These pieces travel with the river until they eventually are deposited in a shallow sea far downstream. This process, called **sedimentation,** continues as other rocks are broken up and also sent downstream. Additional layers of sediment build up over time and bury the original rock particles deeper and deeper, until they become compacted or cemented together. The compacted layers eventually form **sedimentary rock.**

130 HOLT SCIENCE AND TECHNOLOGY INTERACTIVE EXPLORATIONS TEACHER'S GUIDE

Stop 2
As more layers pile up due to continued sedimentation, the sedimentary rock gets buried even deeper within the Earth, and the weight of the layers above it increases. Heat and pressure build under the weight, causing subtle changes in the molecular arrangement of the sedimentary rock. The molecules in the mineral grains of the rock rearrange themselves. This process, called **metamorphosis,** causes chemical changes, and eventually a **metamorphic rock** is born.

Stop 3
As more time passes, the Earth's landmasses rearrange themselves through earthquakes, volcanoes, and other instances of plate movement. Tremendous forces push the metamorphic rock deeper and deeper into the Earth's crust, where intense heat and pressure from the Earth's interior begin to melt the metamorphic rock. As the rock melts, it forms magma, which rises upward through rocks of greater density. Some of this magma eventually bubbles to the surface through a volcano.

Stop 4
The volcano erupts and spews molten rock into the air, which flows as lava down the side of a mountain. The lava cools and forms **igneous rock.**

The Cycle Continues . . .
Immediately, wind, rain, and frost begin to wear away the new igneous rock, turning it into tiny particles that wash into a nearby stream, and the cycle begins again.

Rock On!

Igneous Rocks
Igneous rocks are produced by the cooling of molten rock either at or below the Earth's surface. If the rocks form from magma below the Earth's surface, then they are called **intrusive igneous rocks.** These rocks take a long time to cool and solidify slowly because they are deep underground. This allows time for large crystals to grow. If igneous rocks are formed from lava on the Earth's surface, then they are called **extrusive igneous rocks.** These rocks cool quickly and, as a result, do not have time to form large crystals. They have either very fine crystals or no visible crystals at all.

Most igneous rocks are made up of only a few elements, including silicon, aluminum, magnesium, calcium, and iron. A rock's mineral content affects its color and density. Igneous rocks that are dark in color often contain minerals rich in iron and magnesium, which make the rocks dense and heavy. Igneous rocks that are light in color usually contain calcium and sodium, which make them less dense than the darker igneous rocks.

Sedimentary Rocks
Sedimentary rocks may be made up of different sized particles and sediments, with or without fossils, or they may form from dissolved particles in solutions.

Clastic sedimentary rocks are made of broken pieces of rock. In fact, the word *clastic* comes from the Greek word for *broken.* The pieces of rock that form clastic sedimentary rocks can vary in size from silt to grains of sand to pebbles and larger particles. Transportation of the particles by wind and water sorts them by size. Larger particles, like pebbles, are transported by swift-moving water, while smaller particles, such as silt and mud, are transported by slow-moving water. Clastic rocks can be coarse-grained, small-grained, or fine-grained and may be composed of sharp, angular, or smooth and rounded fragments.

Chemical sedimentary rocks form when minerals separate from solutions. When water evaporates from solutions and the dissolved minerals crystallize, rocks form. Under specific conditions, other minerals *precipitate* (or fall out of solutions) to form rocks.

Organic sedimentary rocks are formed from the remains of once-living organisms. These rocks can form when marine animal shells and skeletons settle on the ocean bottom. Coal is another example of an organic sedimentary rock. It forms on and beneath the Earth's surface when masses of dead plants become extremely compressed over millions of years.

Exploration 6 CD-ROM articles, continued

Metamorphic Rocks

Metamorphic rocks are classified according to structural or chemical changes that result from the rearrangement of the crystals in the rock. Some metamorphic rocks, called **foliated metamorphic rocks,** have visible bands of minerals, and others, called **nonfoliated metamorphic rocks,** do not. Some foliated rocks consist of fine crystals and have very little foliation. Others have coarse grains, and the foliation is obvious. The size of the crystals is an indication of the severity of the temperature and pressure the rocks were exposed to. High temperatures and high pressure produce definite banding in metamorphic rocks.

A Geologic Legacy

Yosemite National Park

Along California's eastern border lies Yosemite National Park—an area famous for its breathtaking vistas, waterfalls, and giant redwood trees. The park also contains part of the Sierra Nevada mountain range. The amazing landscape gives insight into the geological history of this area. Over 500 million years ago, a sea covered the area that is now California. Sediments were deposited on the sea floor in layers, and eventually these layers became sedimentary rock. Tectonic forces lifted the sedimentary rock above sea level, ultimately forming mountains. Far underground, slowly cooling magma rose to the surface. The cooled magma formed granite. Over time, erosion of the overlying rock has revealed the granite that can be seen in the park today. Frost, water, and wind wore away at the landscape for many millennia. Then, beginning about 2 million years ago, a series of ice ages occurred. During each glacial episode, the land became covered with ice. The massive, grinding weight of moving ice, thousands of meters thick, gouged out deep valleys and rounded the mountains' tops. The result is the classic glacial landscape we see today, with deep, U-shaped valleys, waterfalls, and beautifully sculpted landforms.

Exploration 7
Teacher's Notes

Space Case

Key Concepts	A scientific model is a good way to represent or approximate an actual scientific phenomenon. The Moon goes through eight distinct phases as it orbits the Earth.
Summary	Estelle de la Luna is the director of Desert View Park. She has sent a piece of equipment to Dr. Labcoat's lab that was donated to the park. She wants to know what the equipment is for and how park visitors can use it when they go to the visitors' center.
Mission	Describe how best to use some equipment to demonstrate astronomical phenomena.
Solution	The equipment is a planetarium model that should be used to demonstrate the lunar cycle. From the perspective of the Earth in the model, visitors can see each of the Moon's eight phases by looking at the Moon as it revolves around the Earth. An overhead view will demonstrate how one-half of the Moon's surface is lit by the Sun at all times.
Background	Scientific study of the lunar cycle has led to some interesting facts about the Earth's past. In the 1700s, scientists and philosophers noticed discrepancies between the calculated and actual positions of solar eclipses. These discrepancies are due to the Moon's gravity and its effects on the Earth's rotation. Ocean tides on Earth are created by the gravitational attraction between the Moon and the Earth. This gravitational attraction creates tidal friction that slows the Earth's rotation. In our present-day perspective, the slowing rate is minimal, only about 0.00002 seconds per year. However, it has had a profound effect on the length of the Earth's day and year and the lunar month. Today the lunar month is about 27 days long. During the Ordovician period, 420 million years ago, the lunar month was 10 days long. Because the Earth and the Moon revolve around a common center of gravity, and because the angular momentum of the two bodies (the product of their masses and velocities) remains constant, a shorter lunar month implies that the Moon once was much closer to the Earth and that the Earth had once rotated much faster, resulting in a shorter day. As the Earth's rotation slows down over time, the Moon gains momentum and slowly withdraws from Earth, increasing the length of the lunar month.

Exploration 7 Teacher's Notes, continued

Teaching Strategies

Successful completion of this Exploration depends on students developing and maintaining an understanding of what the model demonstrates. For example, the positions are numbered in this particular order because the lunar cycle is usually discussed in terms of starting and ending at new Moon. The Moon does go through intermediate phases or positions—between position 1 and position 2, for example—but the eight phases shown are the most recognized and distinct. You may wish to explain to students that reproducing a three-dimensional model on a two-dimensional computer screen has its limitations. The overhead view of the model in the lab cannot illustrate the 5-degree tilt from the plane of the ecliptic that the Moon's orbit follows. This concept will be useful for students to remember when considering the frequency of solar and lunar eclipses.

As an extension of this Exploration, you may want to have students make their own astronomical models based on the one used in this Exploration. You may decide to incorporate the Moon's 5-degree tilt from the plane of the ecliptic and have students demonstrate solar and lunar eclipses.

Bibliography for Teachers

Bruning, David. "Students in Cyberspace." *Astronomy,* 23 (10): October 1995, p. 48.

Hatchett, Clint. *Discover Planetwatch.* Hyperion, 1993.

Sawicki, Mikolaj. "Using the Solar Eclipse to Estimate Earth's Distance From the Moon." *The Physics Teacher,* 34 (4): April 1996, p. 232.

Bibliography for Students

Brenner, Barbara. *Planetarium.* Bank Street Museum Book, Bantam Books, 1993.

Shibley, John. "Annulus Americanus." *Astronomy,* 22 (9): September 1994, p. 80.

Other Media

Distant Suns—First Light 1.0
CD-ROM
Virtual Reality
2341 Ganador Court
San Luis Obispo, CA 93401
805-545-8575
800-829-8754

In addition to the above CD-ROM, students may find relevant information about the cycles of the Moon and other aspects of astronomy by exploring the Internet. Interested students can search for articles using keywords such as *astronomy, Moon, lunar phases, lunar eclipses,* and *planetariums.*

Name _____ Date _____ Class _____

Exploration 7
Worksheet

Space Case

1. You've been asked to help Estelle de la Luna. What is your mission?

2. Describe the different parts of the equipment on the front table in the lab.

3. What happens when you click on the equipment? Describe the setup now.

4. What do you think the numbered squares represent?

EXPLORATION 7 • SPACE CASE   135

Name _____ Date _____ Class _____

Exploration 7 Worksheet, continued

5. Why do you think the squares are numbered counterclockwise?

_____

_____

_____

6. In the following chart, describe what the Moon looks like at each numbered location that you click:

Position number	Description of Moon

Name _____ Date _____ Class _____

Exploration 7 Worksheet, continued

7. Dr. Labcoat wants you to identify the name of each of the numbered positions on the model. What are they? (Hint: Check out the CD-ROM articles.)

   _____
   _____
   _____
   _____
   _____

8. What should you tell visitors about the scale of this model?

   _____
   _____
   _____
   _____
   _____
   _____
   _____
   _____

9. What other astronomical phenomena could this model demonstrate?

   _____
   _____
   _____
   _____
   _____
   _____

Record your answers in the fax to Ms. de la Luna.

EXPLORATION 7 • SPACE CASE

Name _____ Date _____ Class _____

Exploration 7
Fax Form

FAX

To: Estelle de la Luna (FAX 520-555-6666)

From:

Date:

Subject: Space Case

What does the planetarium model demonstrate?
_____
_____
_____

Determine the correct name for each of the eight numbered positions on the model.

	Position 1	Position 2	Position 3	Position 4	Position 5	Position 6	Position 7	Position 8
first quarter								
full Moon								
last quarter								
new Moon								
waning crescent								
waning gibbous								
waxing crescent								
waxing gibbous								

138 HOLT SCIENCE AND TECHNOLOGY INTERACTIVE EXPLORATIONS TEACHER'S GUIDE

Name _____ Date _____ Class _____

Exploration 7 Fax Form, continued

Please write brief, easy-to-follow directions explaining how to use this equipment.

_____
_____
_____
_____
_____
_____
_____
_____
_____
_____
_____
_____
_____
_____
_____
_____
_____
_____

How much of the Moon's surface is sunlit during each of the eight phases?

_____
_____
_____
_____

Space Case

The following articles can also be found by clicking the computer in the CD-ROM laboratory for Exploration 7:

- *A Hole in the Sky*
- *Phases of the Moon*
- *Moon Science*

Exploration 7
CD-ROM Articles

A solar eclipse

A Hole in the Sky

What Is an Eclipse?

An **eclipse** occurs when, from a certain frame of reference, one celestial object passes in front of another. A **solar eclipse** occurs when the Moon passes between the Earth and the Sun. So the Moon eclipses, or covers, the Sun. During a total solar eclipse, the disk of the Moon obscures the Sun, leaving visible only the Sun's radiant corona. During a partial solar eclipse, the Moon appears to take a bite out of the Sun, covering only part of it. A **lunar eclipse** occurs when the Earth passes between the Moon and the Sun. During a lunar eclipse, the shadow of the Earth is cast across the face of the Moon, leaving the Moon only dimly visible to Earthbound observers. In order for eclipses to occur, the Sun, Moon, and Earth have to line up in the same orbital plane. Since the Moon's orbit around the Earth is tilted 5 degrees from the plane of the ecliptic, eclipses do not occur every time the Moon orbits the Earth.

Solar Eclipse

Few natural events are as spectacular as a solar eclipse, but this rare event is actually an optical illusion. From Earth, the Sun and Moon appear to be about the same size. In actuality, the Sun's diameter is about 400 times greater than the Moon's diameter. But because the Sun is also 400 times farther away from the Earth than is the Moon, the Sun and Moon appear to be about the same size in the sky. Thus, during a total solar eclipse, when the Moon moves between the Earth and Sun, it appears to completely cover the disk of the Sun.

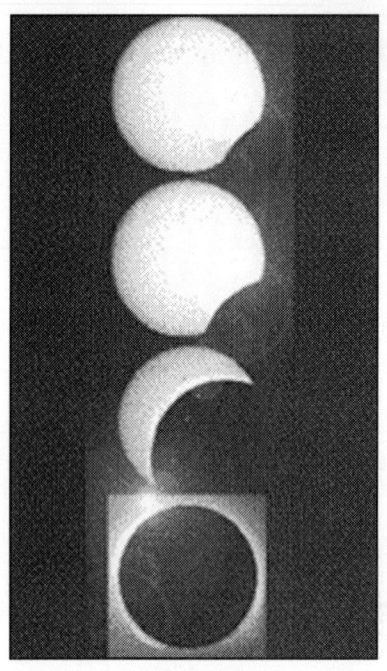

Lunar Eclipse

During a lunar eclipse, the Earth lines up between the Moon and the Sun, and the Moon moves into the shadow cast by the Earth. Although the Earth's shadow is large enough to completely cover the Moon, the Moon usually remains dimly visible because of light scattered by the Earth's atmosphere. If you watch a lunar eclipse, you can actually see the shadow of the Earth move across the Moon's surface. From observations of this movement, early astronomers deduced that the Earth was a sphere.

Phases of the Moon

The Sunlit Moon

Like the Earth, half of the Moon is always lit and half is always in darkness (except during lunar eclipses, of course). However, our view of the Moon changes as the Moon travels around the Earth. How much of the Moon's sunlit face we see depends on the relative positions of the Moon, Earth, and Sun. The **phases of the Moon** are determined by how much of the Moon's sunlit half is facing the Earth.

140   HOLT SCIENCE AND TECHNOLOGY INTERACTIVE EXPLORATIONS TEACHER'S GUIDE

New Moon
When the Moon's dark side faces the Earth, the Moon is in the new phase. This happens when the Moon is between the Sun and the Earth. Solar eclipses occur only during a new Moon.

Waxing Crescent Moon
The word *waxing* means to increase or get stronger. After the new Moon, a sliver, or crescent, of the Moon appears in the sky. This phase results as the Moon rotates counterclockwise around the Earth and a tiny portion, or crescent, of its surface becomes visible from Earth. This crescent is due to light from the Sun.

First Quarter Moon
The first quarter phase allows us to see one-fourth of the sunlit side of the Moon. It looks like a half-circle in the sky. First quarter Moon occurs when the Earth forms a 90-degree angle with the Sun and the Moon.

Waxing Gibbous Moon
The word *gibbous* comes from the Latin word for "hump." Notice how the gibbous Moon looks almost circular, like a hump. The waxing gibbous Moon resembles a circle because most of the surface that is facing the Earth is lit by the Sun.

Full Moon
When the entire sunlit side of the Moon faces the Earth, the Moon is full. Full Moon occurs when the Earth is between the Sun and the Moon. Lunar eclipses occur only during the full Moon phase.

Waning Gibbous Moon
The word *waning* means to decrease or get smaller. After the full Moon, the waning gibbous Moon appears. Notice how the portion of the Moon lit during waning gibbous Moon is the opposite of the portion lit during waxing gibbous Moon.

Last Quarter Moon
Compare the last quarter of the Moon's phase with the first quarter. As with first quarter Moon, last quarter Moon also occurs when the Earth forms a 90-degree angle with the Sun and Moon. However, during the last quarter, the Moon is on the side of Earth that is opposite the side the Moon was on in the first quarter. Although we see a half-circle in both phases, a different part of the Moon is visible in each quarter.

Waning Crescent Moon
As the cycle of the Moon's phases nears completion, the waning crescent Moon appears. During the waning crescent phase, the lit side of the Moon is lit that is opposite the side lit during the waxing crescent phase.

Moon Science

Moon Facts

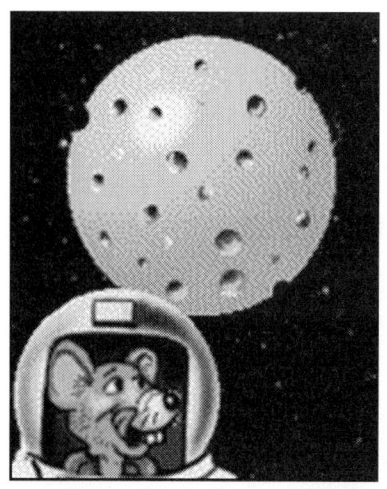

Is the Moon made of green cheese?

Questions lead scientists to make great discoveries. Scientists decided that the best way to find the answers to questions about the Moon was to actually go to the Moon! Developing space technology and putting a person on the Moon have allowed us to learn more about what the Moon is made of as well as some other pretty interesting Moon facts. Check these out:

- The Moon isn't made of green cheese, but it does have a rough surface. Meteorites striking the Moon have formed millions of craters—some measuring up to 1200 km in diameter.

- The Moon is airless, waterless, and lifeless.

- The Moon is about 384,000 km from the Earth.

- The Moon revolves around the Earth at an average speed of about 3700 km per hour.

- A person who weighs 600 N on Earth would only weigh about 100 N on the Moon! That's because the gravitational force from the Moon is about one-sixth as strong as the gravitational force on Earth.

Scientists also have several hypotheses about how the Moon formed. Here are a few:

- The Moon is a piece of the Earth that broke off when the Earth was still molten material.
- The Moon is a piece of the Earth that broke off in a collision with an asteroid.
- The Moon formed from pieces left over after the Earth formed.
- The Earth's gravitational attraction captured the Moon as the Moon drifted through the solar system.

Sister Moon

The idea of a man in the Moon figures into many children's stories. In fact, many cultures have stories and legends about the Moon. For example, several American Indian traditions recognize the Moon as a sister to the sky. In addition, American Indian legends consider the Moon an important force in the origin of the world. Greek and Roman mythologies both had a Moon goddess who was known by several names—Artemis, Diana, and Luna.

The Moon was once believed to cause a form of insanity—lunacy. It supposedly brought out werewolves and other monsters. A mild form of lunacy was thought to make people fall madly in love with each other. Even today, the Moon is often featured in romantic songs and poems.

One Small Step . . .

On July 20, 1969, Neil Armstrong became the first human to walk on the Moon.

The race to land a human on the Moon began in the 1940s. For about 20 years, the United States and the former Soviet Union competed for this distinction. In July of 1969, the United States won the race. Containing astronauts Neil Armstrong and Edwin "Buzz" Aldrin, the Lunar Lander from *Apollo 11* landed successfully on the Moon. As Neil Armstrong stepped off the ladder onto the Moon, he uttered the now-famous statement, "That's one small step for man, one giant leap for mankind."

How's It Growing?

Key Concepts	Conducting a controlled experiment is a good way to deduce explanations for observed phenomena. A hydrangea plant can produce either blue or pink flowers, depending on the pH of the soil.
Summary	Ms. Rosie Flores is trying to answer a gardening question for a newspaper column. A reader has had some problems with his new hydrangea plants. Unlike his other hydrangeas, which produce blue flowers, these new plants, although genetically identical to the others, produce pink flowers. Ms. Flores has sent two of these new plants to Dr. Labcoat's lab to be examined. She wants to know why they are producing pink flowers instead of blue ones.
Mission	Unearth the answer to a colorful floral mystery.
Solution	The new hydrangea plants are producing pink flowers as opposed to blue ones because the soil in which the new plants are growing is more alkaline than that of the blue-flowering hydrangeas. Adding soil conditioner makes the soil more acidic, and as a result the new hydrangea plants produce blue flowers.
Background	Hydrangea plants are lush flowering shrubs that may be grown indoors or outdoors. Some outdoor varieties can grow as high as 8 m. Hydrangeas require moist, fertile soil with large amounts of organic matter. They grow well in shaded or partly shaded environments and tend to dry up when exposed to excessive sunlight. Hydrangeas are popular plants because of their spectacular blooming in late summer and autumn.
	Hydrangea macrophylla is a species of spreading shrub that can produce either blue or pink flowers. The different colors are demonstrative of the plant's sensitivity to soil pH. Acidic soil results in blue flowers, and alkaline soil results in pink flowers. Because this species spreads horizontally, some of the roots may grow in acidic soil, while others grow in alkaline soil. This allows both colors of flowers to be present on the same plant. Many gardeners and flower enthusiasts enjoy growing hydrangeas because of their ability to control the color of the flowers. Adding lime to the soil around the base of the plant will cause the plant to produce pink flowers, whereas adding aluminum sulfate, called powdered bauxite, to the soil will result in blue flowers.

Exploration 8 Teacher's Notes, continued

Teaching Strategies

To help students complete this Exploration efficiently, emphasize the importance of changing only one variable at a time on the experimental plant. You may wish to suggest that students set up their control plant with the lowest settings of light and with no additional factors. Such a simple setup will make selecting different variables easier. Make sure that students research the CD-ROM articles thoroughly so that they understand the uses of plant food, soil conditioners, and non-chemical pesticides, such as ladybugs. Have them keep a good record of their notes in the Notepad.

As an extension of this Exploration, you could conduct a real-life experiment by growing hydrangeas in your classroom. Make sure to provide the plants with the proper environment—partial shade, plenty of water, and warm temperatures ranging from 18°C to 28°C. Growing two separate plants in two soils of different pH levels will allow students to see exactly how the color of hydrangea flowers are governed by soil pH.

Bibliography for Teachers

Cullen, Kathleen. "Hydrangeas—So Passé, So Today." *American Horticulturist,* 73 (6): June 1994, pp. 25–29.

Lawson-Hall, Toni, and Brian Rothera. *Hydrangeas; A Gardeners' Guide.* London: B.T. Batsford Ltd., 1995.

Bibliography for Students

Attenborough, David. *The Private Life of Plants.* Princeton: Princeton University Press, 1995.

Stewart, Doug. "Luck Be a Ladybug." *National Wildlife,* 32 (4): June/July 1994, pp. 30–32.

Other Media

Garden Encyclopedia 2.0
CD-ROM
Books That Work, Inc.
P. O. Box 3201
Salinas, CA 93912-9869
800-242-4546

In addition to the above CD-ROM, students may find relevant information about hydrangeas and gardening on the Internet. Interested students can search for articles using keywords such as *hydrangeas, gardening tips, botany,* and *horticulture.*

Name _____ Date _____ Class _____

**Exploration 8
Worksheet**

How's It Growing?

1. Rosie Flores needs your gardening expertise. What does she want to know?

   _____
   _____
   _____

2. How did the reader that Ms. Flores is responding to grow his hydrangeas? Describe this process. (Hint: Check out the CD-ROM articles.)

   _____
   _____
   _____
   _____

3. Dr. Labcoat has set out some materials on the front table in the lab. Describe her setup.

   _____
   _____
   _____
   _____

4. Why is there a control plant and an experimental plant?

   _____
   _____
   _____

5. Record the settings you choose for your control setup.

   _____
   _____
   _____

EXPLORATION 8 • HOW'S IT GROWING?   145

Name _____ Date _____ Class _____

Exploration 8 Worksheet, continued

6. Conduct all of the necessary experiments, recording your observations of how each of the variables below affects the hydrangeas.

   a. Hours of light per day

   _____

   _____

   _____

   _____

   b. Soil conditioner

   _____

   _____

   _____

   _____

   c. Plant food

   _____

   _____

   _____

   _____

   d. Ladybugs

   _____

   _____

   _____

   _____

Record your answers in the fax to Ms. Flores.

Name _____ Date _____ Class _____

Exploration 8
Fax Form

FAX

To: Ms. Rosie Flores (FAX 213-555-0612)

From:

Date:

Subject: How's it growing?

How can two hydrangea plants that are genetically identical produce different-colored flowers?

_____

_____

_____

_____

For Internal Use Only

Please answer the following questions for my laboratory records. Scientists must always keep good records *Dr. Crystal Labcoat*

During your experiments, which one of the following variables helped you to discover the correct answer to the above question?

	CHANGE HOURS OF LIGHT PER DAY.
	ADD SOIL CONDITIONER.
	ADD PLANT FOOD.
	INTRODUCE LADYBUGS.

What effects did each of the following variables have on the hydrangeas? Please explain.

a. Hours of light per day

_____

_____

_____

_____

DISC 2

EXPLORATION 8 • HOW'S IT GROWING? 147

Name _____ Date _____ Class _____

Exploration 8 Fax Form, continued

b. Soil conditioner

_____

_____

_____

_____

c. Plant food

_____

_____

_____

_____

d. Ladybugs

_____

_____

_____

_____

> **How's It Growing?**
>
> The following articles can also be found by clicking the computer in the CD-ROM laboratory for Exploration 8:
>
> • *The Dirt on Soil*
> • *How Does Your Garden Grow?*

Exploration 8
CD-ROM Articles

The Dirt on Soil

What Makes Up Soil?

Soil is more than just dirt. It is made up of a mixture of things—nonliving, living, and once-living parts. The nonliving parts of soil include water, sand, clay, silt, broken rocks, and pebbles. The living parts of soil include bacteria and other microorganisms, worms, insects, and plants. When living plants and animals die, their remains decay to form the once-living part of soil. This once-living part of soil is called **humus.**

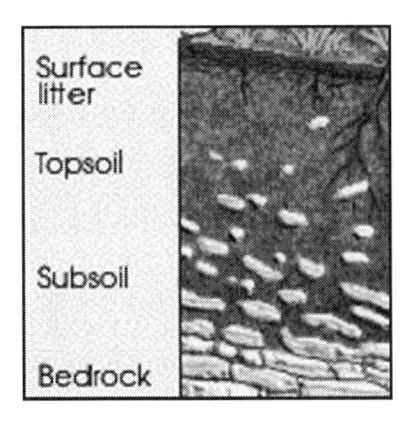

Erosion, weathering, and the breakdown of living things cause soil to change over time. You can see changes in soil by observing soil layers. The very thin top layer is called the **surface litter.** Animal wastes, leaves, twigs, and many living things are part of this layer. Underneath the surface litter is the **topsoil.** Most of the soil's humus is found in this layer. The roots of most plants pass through these two layers. The **subsoil** lies beneath the topsoil. The subsoil is mostly clay, silt, weathered rock, and other nonliving particles. The bottom layer of soil is called **bedrock.** This layer is unweathered rock.

Soil's Chemical Composition

Healthy soils are the beginnings of successful gardens.

Different chemicals in soil affect how acidic or alkaline the soil is. An acidic soil has a relatively low pH, whereas an alkaline soil has a relatively high pH. Some plants will only grow in soil with a certain pH. A change in pH can alter the size, color, and health of a plant.

Soil pH affects the solubility of different minerals in the soil. Because plants differ in the amounts of minerals they need, pH affects the kinds of plants found in soils. For example, raising soil pH increases the solubility of calcium and lowers the solubility of iron. Plants in this soil would be able to absorb more calcium but not as much iron. Alfalfa and sweet clover require a great deal of calcium, so they grow well in soils with a high pH. Azaleas and rhododendrons, two common garden plants, need a great deal of iron, so they grow best in more acidic soils.

Sometimes chemicals are added to the soil to help plants grow. Called *fertilizers,* these chemicals usually add nitrogen or phosphorus to nutrient-poor soils.

How Does Your Garden Grow?

Planning Steps for a Grand Garden

A backyard garden can yield beautiful flowers as well as fresh fruits and vegetables. All gardens, however, need special care and maintenance to yield the healthiest plants.

EXPLORATION 8 • HOW'S IT GROWING? **149**

Exploration 8 CD-ROM articles, continued

Step 1: Choose the location.
A successful garden requires good water drainage, so avoid places that turn swampy after a rain. The ground should be even or slightly sloped; avoid steep hills. Be sure that your garden receives the right amount of light for the kinds of plants you wish to grow. A cactus, for example, needs a lot of direct, bright sunlight. So you would do well to plant your cactus in a well-exposed area that receives little shade throughout the day.

Step 2: Check the soil.
To ensure the health of your plants, check the pH and nutrient content of your garden's soil. A soil test kit and advice from an experienced gardener can help you figure out if your garden's soil needs fertilizer, a soil conditioner, or other additions.

Simple soil tests help gardeners determine the pH of the soil.

Fertilizers add essential nutrients and minerals to soil. The most important nutrients for plants are nitrogen, phosphorus, and potassium, all of which are absorbed by plants from the soil. Manure and compost are inexpensive fertilizers that can make soil more fertile. You can buy fertilizers in a gardening store, or you can make compost yourself out of yard waste and plant scraps.

Soil conditioners are sometimes used along with fertilizers. These additives alter pH, improve the soil's ability to hold water, and improve drainage. Aluminum sulfate, also called powdered bauxite, lowers the pH of soil. Calcium carbonate, or lime, raises the pH of soil. Lime also causes chemical reactions in the soil that make nutrients and minerals more soluble. The result is increased absorption of some nutrients by the plant.

Step 3: Choose the plants.
The best plants for an outdoor garden are plants that naturally grow well in your area and in your type of soil. You can talk to people in a gardening store, or you can get advice from the library about plants that are native to your area. You can also start a garden from existing plants. In a process called **vegetative reproduction,** new plants are grown from stem or leaf cuttings of other plants. This procedure involves planting the cuttings in fertile soil so that they will grow new roots. These new roots can then be replanted in a larger pot, or you can transplant them to your garden.

Step 4: Make a plan.
Once you've chosen your plants, decide how much room each plant will need. If you are growing flowers, imagine what the flower beds will look like once the plants are in bloom. Consider how you want to arrange your garden. Careful planning allows you to decorate the garden with clusters of colors. If you are planting a vegetable garden, find out which plants will be ready to harvest first. Make plans to plant additional vegetables in place of those you have already harvested.

Raising Healthy Hydrangeas
A healthy hydrangea produces bunches of flowers that bloom in a variety of colors, based on the chemical composition of the soil. A hydrangea's flowers may be blue, purple, red, pink, white, or any shade in between. A mature plant may reach over 1 m in diameter, so the shrubs should be widely spaced when planted. A single cluster of flowers may be 20–30 cm in diameter. During the spring, old flowers and stems should be pruned away.

Hydrangeas grow best in semi-shade, and they require a great deal of water. Avoid planting the shrub in full sunlight. Adding mulch to the soil keeps roots moist in the summer and protects the plant from freezing in the winter. The best temperature for growth is about 20–22°C during the day and 13–16°C at night.

Hydrangeas can be grown from seeds or from cuttings. The seeds take two to three months to mature, and then about two weeks to germinate. Tip-cuttings produce plants much more quickly. To produce a plant this way, you should remove a 7–12 cm section from the top of a stem. Then replant this portion in proper soil, light, and temperature conditions.

Pollution-Free Pest Control

Hungry plant pests can turn a beautiful garden into a big mess. While many chemical pesticides do kill pests, they may also harm beneficial insects and pollute the soil and water. Luckily, there are alternatives to chemical pesticides. Here are a few non-chemical ways to prevent or limit damage from pests.

- Regularly remove weeds and plant scraps, such as dead leaves. These items are breeding grounds for pests.

- Introduce natural predators that can control pests biologically. Ladybugs, for example, eat aphids, which can damage many plants.

- Use organic pesticides. Some plants naturally produce chemicals to protect themselves from pests. These chemicals can be extracted from plants and sold as organic pesticides. Not only do these products keep some pests away, but they also break down into harmless substances.

Exploration 1
Teacher's Notes

The Nose Knows

Key Concepts — Diffusion is the process in which particles move from an area of higher concentration to an area of lower concentration. The sense of smell can be used to detect warning odors and to protect humans from danger.

Summary — Ms. Dee Foushen is the director of a school for the hearing- and sight-impaired. She needs help designing a special fire alarm to ensure the safety of her students. Since some of her students cannot see warning lights or hear a typical alarm, she wants to install an "odor alarm" that would release an odorous chemical in the building in the event of a fire. Ms. Foushen has sent five samples of odorous chemicals to Dr. Labcoat's lab and wants to know which one would be the best choice.

Mission — Choose the best chemical to use for an odor alarm.

Solution — Cinnamon is the best odorous chemical to use for an odor alarm because its scent diffuses fairly quickly through the air and it is not a dangerous substance. Rotten eggs and alcohol both diffuse more quickly than cinnamon does, but these chemicals could be dangerous to students if used for the odor alarm.

Background — Although humans rely primarily on sight and sound to gather information, the human sense of smell has some interesting functions. Because of the anatomical structure of the brain, the sense of smell is closely associated with memory. Information from other sensory neurons (sight, touch, and hearing) gets routed through the thalamus only; information from olfactory neurons (including those that stimulate taste) goes to the thalamus as well as to portions of the brain associated with memory, namely the hippocampus (short-term memory) and the amygdala (long-term memory). This makes the sense of smell a powerful memory stimulator. The connection between smell and memory explains why some odors immediately trigger memories of a person, place, or time in the past.

Scientists are exploring the connection between smell and memory in research for Alzheimer's disease. This neurological disorder causes interruptions in the transmission of nerve impulses across the synapses in the brain. Alzheimer's causes concentrated synaptic loss in the limbic system, where the hippocampus and amygdala are located. As a result, many people with progressive Alzheimer's suffer from memory loss as well as a diminished sensitivity to smell, a condition called anosmia. Because of the difficulty in diagnosing Alzheimer's, some scientists have suggested monitoring declines in sensitivity to smell as a way to help identify the early stages of Alzheimer's.

EXPLORATION 1 • THE NOSE KNOWS

Exploration 1 Teacher's Notes, continued

Teaching Strategies

Students may be tempted to choose a substance for Ms. Foushen's odor alarm based only on its rate of diffusion. Emphasize that this is not the best approach, as it may lead students to choose an unsafe substance. For example, rotten eggs contain hydrogen sulfide, a toxic chemical, and alcohol is flammable. Encourage students to research each substance in the CD-ROM articles before making a decision.

As an extension of this Exploration, you may want to conduct a classroom activity such as the following: Set up an odor kit of various unknown odors (available through biological supply companies), and have students time how long it takes to sense each odor as the smell particles diffuse through the air. Students could then recommend an odor for use in an odor alarm. You can also ask students about ways to increase the rate of diffusion of each odor. For example, substances diffuse faster at higher temperatures.

You may also wish to discuss with students the possible limitations of implementing an odor alarm in Ms. Foushen's school for the hearing- and sight-impaired. Questions to ask students might consist of the following: What kind of sensor would detect the fire? How would the cinnamon be stored and released? Would the cinnamon be messy? Would it stain clothing or cause allergic reactions? If someone brought a fresh-baked cinnamon roll into the building, would it be mistaken for the odor alarm? Can you think of any other ways to notify the students of an unsafe situation?

Bibliography for Teachers

Taubes, Gary. "The Electronic Nose." *Discover,* 17 (9): September 1996, p. 40.

Trum Hunter, Beatrice. "The Sales Appeal of Scents." *Consumers' Research Magazine,* 78 (10): April 1995, p. 48.

Bibliography for Students

Lipkin, Richard. "Tracking an Undersea Scent." *Science News,* 147 (5): February 4, 1995, p. 78.

Schwenk, Kurt. "The Serpent's Tongue." *Natural History,* 104 (4): April 1995, p. 48.

Other Media

Diffusion and Osmosis
Videotape
Encyclopædia Britannica Educational Corporation
310 S. Michigan Ave.
Chicago, IL 60604-9839
800-554-9862

In addition to the above video, students may find relevant information about diffusion by exploring the Internet. Interested students can search for articles using keywords such as *diffusion, osmosis,* and *semipermeable membranes.* Students can also access information about the *sense of smell* by exploring the Internet.

Name _____ Date _____ Class _____

Exploration 1 Worksheet

The Nose Knows

1. Ms. Foushen needs your help sniffing out the solution to a problem. What has she asked you to do?

2. Explain the process of diffusion. (Hint: Check out the wall chart in the lab.)

3. What is the difference between diffusion and osmosis? (If you're not sure, check out the CD-ROM articles.)

4. What is the equipment on the front table in Dr. Labcoat's lab designed to do?

EXPLORATION 1 • THE NOSE KNOWS   155

Name _____ Date _____ Class _____

Exploration 1 Worksheet, continued

5. Use the equipment to conduct the necessary tests, and record your data in the table below.

Test tube	Test-tube contents	Time to diffuse (sec.)
A	perfume	
B	rotten eggs	
C	garlic	
D	alcohol	
E	cinnamon	

6. Why are the temperature and pressure kept constant for this experiment? (If you're not sure, check out the CD-ROM articles.)

_____
_____
_____
_____

7. What does the equipment on the back counter in Dr. Labcoat's lab demonstrate?

_____
_____
_____
_____

8. How do you smell an odorous chemical? Use the CD-ROM articles to help you explain how your sense of smell works.

_____
_____
_____
_____
_____

Record your answers in the fax to Ms. Foushen.

Name _____ Date _____ Class _____

Exploration 1
Fax Form

FAX

To: Ms. Dee Foushen (FAX 512-555-7003)

From:

Date:

Subject: The Nose Knows

Which of the five samples do you recommend that I use for the fire alarm?

| Alcohol | Cinnamon | Garlic | Perfume | Rotten eggs |

Please explain why you chose this sample.

_____
_____
_____
_____
_____
_____

Explain how odors spread through a room.

_____
_____
_____
_____
_____
_____

> **Exploration 1**
> **CD-ROM Articles**

> ### The Nose Knows
> The following articles can also be found by clicking the computer in the CD-ROM laboratory for Exploration 1:
> - *Diffusing the Confusion*
> - *Making Sense of Scents*

Diffusing the Confusion

What Is Diffusion?
All matter is made up of particles that are in constant motion. Even the air around us consists of billions of particles that are moving at high speeds in random directions. This characteristic of matter allows diffusion to take place.
Diffusion is the process in which particles of one substance move from an area of higher concentration to areas of lower concentration.

The particles of food coloring are moving from an area of higher concentration to areas of lower concentration.

How does this work? Look at the photo showing what happens when a drop of food coloring is added to water. The food coloring is an area of high concentration (of food coloring particles), and the water is an area of low concentration (of food coloring particles). As soon as the drop of food coloring hits the water, the particles of food coloring immediately move toward areas of lower concentration in the water. After a period of time, the particles reach a relatively uniform concentration, and diffusion ceases.

Snowball Diffusion
To make diffusion easier to understand, imagine that Alan and Byron are having a snowball fight across the fence that separates their backyards. Each boy is allowed to throw a set number of snowballs, say one-tenth of them, into the other boy's yard every minute.

Suppose that at the beginning of the fight Alan has 100 snowballs in his yard and Byron has 50 snowballs in his yard. Alan's yard is the area of higher concentration, and Byron's yard is the area of lower concentration. In the first minute, Alan throws 10 snowballs (one-tenth of 100) into Byron's yard and Byron throws 5 snowballs (one-tenth of 50) into Alan's yard. At the end of the first minute, Alan has 95 snowballs in his yard and Byron has 55 snowballs in his yard. Alan has a decreased concentration of snowballs, and Byron has an increased concentration of snowballs. Therefore, there is a net movement of snowballs from the area of higher concentration to the area of lower concentration.

During the second minute, Alan again throws one-tenth of his snowballs into Byron's yard and Byron again throws one-tenth of his snowballs into Alan's yard. At the end of the second minute, Alan has 91 snowballs in his yard and Byron has 59 snowballs in his yard. As time passes, Alan loses snowballs while Byron gains snowballs because the net movement of snowballs continues to be from Alan's yard to Byron's yard.

As Alan and Byron continue to play, they move toward having an average of 75 snowballs each. At this point, there is a uniform concentration of snowballs in each yard. If they continued playing, on average they would each throw the same number of snowballs every minute. There would be no net movement of the snowballs, and the process of "snowball diffusion" essentially stops.

Different Rates of Diffusion

In general, diffusion occurs more readily between gases and between liquids than it does between solids. This is because the particles that make up gases and liquids are farther apart and move faster than the particles in solids. Although it may be difficult to imagine, diffusion does take place between solids. For example, if samples of zinc and copper are clamped together for several months, a small amount of diffusion between particles of zinc and copper will occur.

Concentration

Differences in concentration affect the rate of diffusion. In general, diffusion occurs more rapidly when there is a large difference in concentration between two areas. Refer back to the example of the food coloring in the beaker of water. The difference in concentration (of food coloring particles) between the drop of food coloring and the water is greatest when the drop of food coloring first enters the water. Initially, diffusion occurs very rapidly. As diffusion continues, the difference in concentration between the water and the food coloring lessens, and diffusion begins to slow down. Once the food coloring particles are uniformly distributed throughout the water and there is little difference in concentration, then diffusion effectively stops.

Temperature

Differences in temperature also affect the rate of diffusion. When the temperature of a substance is raised, the molecules move faster and rebound farther after collisions due to an increased amount of kinetic energy. Consider how much faster a spoonful of sugar dissolves in a cup of hot tea than it does in a glass of iced tea.

Pressure

The rate of diffusion is also affected by pressure. Under high pressure, particles are squeezed closer together. As a result, there is less time between particle collisions, and particles get sent in new directions at a faster rate. This increased rate of collisions spreads the particles out faster, increasing the rate of diffusion.

Osmosis—Diffusion Through a Membrane

In living systems, water particles often diffuse through a semipermeable membrane, such as a cell membrane. A **semipermeable membrane** allows certain particles to pass through it while blocking others. In general, semipermeable membranes prevent the passage of larger particles. The process of diffusion through a semipermeable membrane is called **osmosis.**

The concentration of water inside a cell affects whether water moves into or out of the cell through the cell membrane. When there are more water particles outside the membrane than there are inside the membrane, water moves into the cell. When there are fewer water particles outside the membrane than there are inside the membrane, water moves out of the cell. When the concentration of water particles is equal on either side of the membrane, there is no net movement of water particles in either direction.

Making Sense of Scents

Wow, Do You Smell!

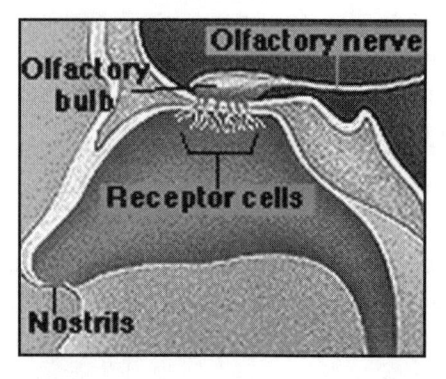

Your olfactory system allows you to recognize a variety of odors—from the aroma of fresh-baked cookies to the stink of rotten garbage. You sense these smells by inhaling particles that have diffused from their source into the air. Receptor cells inside your nose react to these particles by sending a message along the olfactory nerve to the olfactory bulbs in the brain. There, the messages are interpreted into the sensation of smell.

Smelling is not the only purpose your olfactory system serves, however. It also plays an important function in your sense of taste. If you have ever held your nose to eat or drink something you dislike, cold medicine, for instance, then you have experienced first-hand the effects of your olfactory system on your sense of taste. In fact, if you were blindfolded and wearing noseplugs, you probably could not tell an apple from an onion just by tasting them.

It Is a Diffusing World

As you know, many things have a scent, and these scents can provide valuable information. Many animals rely heavily on their sense of smell to find food, shelter, and mates. Some animal smells can also serve as warnings. For example, skunks are famous for their smelly emissions. A skunk will "spray" if it feels threatened. This offensive tactic helps the skunk thwart other animals.

Smell also plays an important role in human behavior. From infancy, you develop definite opinions about different odors. As your brain learns to recognize different odors, it can determine safe smells from dangerous smells. For example, you might not drink sour milk because particles diffusing from the milk let you know that the milk is not safe to drink. In this way, your sense of smell helps you make sensible decisions.

Smelly Chemicals

Most people can recognize thousands of different objects just by their smells. The distinctive scent of a substance is caused by different chemicals or combinations of chemicals.

Food

Certain chemicals give foods their distinctive smells. For example, butyl acetate makes an apple smell like an apple, while ocytl acetate makes an orange smell like an orange. The names of some scent-producing chemicals sound like the substance that produces the odor. For example, cinnamaldehyde is the chemical that gives cinnamon its fiery smell, and vanillin gives the vanilla bean its odor. Other names of smelly chemicals are derived from the scientific name of the substance. For example, the scientific name for garlic is *Allium sativum,* and the chemical that gives garlic its smell is called allicin.

Alcohol

If you open a bottle of rubbing alcohol, you will probably be able to smell the alcohol almost immediately. This is because rubbing alcohol diffuses very quickly. Rubbing alcohol is often used as a solvent because it reacts with many different kinds of chemicals. For example, one way to remove ink from skin and clothing is by applying rubbing alcohol to the stain. Alcohol is also highly flammable. If a flame gets anywhere near the quickly diffusing particles of alcohol, you will witness a sure-fire reaction.

Perfume

Perfumes may contain over 100 different ingredients. The most familiar ingredients come from fragrant plants or flowers, such as sandalwood or roses. Other perfume ingredients come from animals and from human-made chemicals. Some substances, like civet musk, are used to make the odors in the perfume last longer. Some perfumes sold in stores are synthetic, which means that the chemicals were created and mixed in a laboratory. For example, a combination of geraniol and beta-phenyl ethyl alcohol produces a scent that smells like roses. Many perfumes, especially those packaged in spray bottles, contain isopropyl alcohol, which allows the scent to diffuse quickly. However, this ingredient also makes the perfume flammable.

Hydrogen Sulfide

Some dangerous chemicals produce distinct odors. Hydrogen sulfide, for example, is a toxic, gaseous chemical that smells like rotten eggs. The brain recognizes the smell of hydrogen sulfide as unpleasant, and we instinctively want to get away from the smelly source. However, not all dangerous substances produce a smelly warning. Natural gas, for example, has no odor, so gas companies mix odor-causing chemicals with the gas. This way, you can detect natural gas leaks before they reach toxic or explosive levels in your home.

Exploration 2
Teacher's Notes

Sea the Light

Key Concepts	Materials can differ in density because of their structure at the particle level. An underwater lamp that is neutrally buoyant will neither sink nor rise in sea water.
Summary	Diane Sittie, a scuba-diving enthusiast, wants to create an underwater lamp that she can use where she dives. She has sent a prototype lamp base and some ballast disks to Dr. Labcoat's lab. Ms. Sittie needs to know which ballast disk to add to the lamp base so that the entire lamp will be neutrally buoyant in the sea water where Ms. Sittie dives.
Mission	Recommend a metal ballast disk for an underwater hanging lamp.
Solution	A titanium ballast disk used in conjunction with the waterproof lamp base will result in an underwater hanging lamp that has a total density closest to that of the water where Ms. Sittie dives. As a result, the lamp will be neutrally buoyant.
Background	Scuba diving requires special equipment. The most obvious part of the equipment is the breathing apparatus (*scuba* stands for *s*elf-*c*ontained *u*nderwater *b*reathing *a*pparatus). To allow a diver to breathe underwater, a tank of compressed air, usually consisting of a mixture of helium, oxygen, and nitrogen, is attached to hoses and regulators. One regulator is connected to the tank. Called a *first-stage regulator,* it controls how much compressed air flows from the tank through the hoses, and how compressed the air is. This regulator allows a diver to decompress as he or she surfaces, ensuring a safe transition from breathing compressed air to breathing normal air. Another regulator, called a *second-stage regulator,* is attached to the diver's mouthpiece. This regulator controls the opening and closing of a mechanism in the mouthpiece that determines how much force the diver must use to inhale and exhale.
Another important piece of equipment used by scuba divers is a buoyancy control device, or BCD. BCDs are usually similar to vests or backpacks; they are designed to hold the air tank and other accessories. BCDs also allow divers to control their buoyancy underwater. By inflating the BCD, the diver rises toward the surface, and by deflating the BCD, the diver sinks. The inflation and deflation can be done orally or by an automatic inflator. Because the human body tends to float rather than sink, scuba divers in ocean water also use weights as ballast. The weights allow them to descend to certain depths as well as give them stability and mobility underwater. Some BCDs have pockets designed to hold these weights, and some divers wear weight belts.

Exploration 2 Teacher's Notes, continued

Teaching Strategies

Because students may have difficulty connecting the concept of density with the particle theory of matter, you may need to discuss how the atomic models on the back counter of Dr. Labcoat's lab demonstrate density on the particle level. Make sure students understand that these atomic models show how much matter occupies a given space (density) for a given element. You may want to use the periodic table to explain that elements with higher atomic numbers can sometimes also have greater densities because the atomic number indicates the number of particles (protons) within an atom of a given element. You can also use the atomic models to explain how particle arrangement (how close together or far apart particles are within a given molecule) can also determine the density of a substance.

To help students understand how the density of fluids can affect buoyancy, you may want to conduct a demonstration that shows how fluids with greater densities can exert greater buoyant forces. Mix a solution of salt and water that is concentrated but not saturated and that remains clear in color. Pour this solution into a glass or a beaker. Fill an identical glass or beaker with fresh water. Place a whole raw egg in each water sample. The egg placed in the salt water should float, and the egg placed in the fresh water should sink. Explain this phenomenon to students in terms of buoyant force and density.

Bibliography for Teachers

de Grasse Tyson, Neil. "On Being Dense." *Natural History,* 105 (1): January 1996, pp. 66–67.

Peterson, I. "Explosive Expansion of Atomic Nuclei." *Science News,* 147 (15): April 15, 1995, p. 228.

Bibliography for Students

Hoover, Pierce. "The Deep." *Popular Mechanics,* 173 (1): January 1996, p. 68.

Surkiewicz, Joe. "Derby of the Deep." *Boys' Life,* 86 (7): July 1996, p. 42.

Other Media

The Atom
Video and book
SVE (Society for Visual Education)
6677 N. Northwest Highway
Chicago, IL 60631
800-624-1678

In addition to the above video, students may find relevant information about the particle theory of matter, density, and buoyancy by exploring the Internet. Interested students can search for articles under *chemistry* using keywords such as *atoms; density; protons, neutrons, and electrons;* and *atomic models.* Students could also search under *physics* to find information about *buoyancy.* Students can also access information about *scuba diving, scuba gear,* and *scuba certification* on the Internet.

Name _____ Date _____ Class _____

Exploration 2
Worksheet

Sea the Light

1. Ms. Sittie wants to create an underwater hanging lamp. What help does she need from you?

   _____
   _____
   _____
   _____
   _____

2. What does *ballast* mean? (If you aren't sure, use the CD-ROM articles to help you.)

   _____
   _____
   _____
   _____

3. What purpose do you think the ballast disks serve in the design of the underwater lamp?

   _____
   _____
   _____
   _____
   _____

4. Describe how you will use the equipment on the lab's front table to answer Ms. Sittie's questions.

   _____
   _____
   _____
   _____
   _____
   _____
   _____

Name _____ Date _____ Class _____

Exploration 2 Worksheet, continued

5. Use the equipment to conduct all of the necessary tests, and record your data in the first two columns of the table below. Then use your results to calculate the values for the third column.

Ballast disk	Mass (g)	Volume (mL)	Density (g/mL)
A Copper			
B Aluminum			
C Brass			
D Titanium			
E Platinum			
F Zinc			

6. Calculate the total density of the entire lamp for each individual ballast disk. (If you aren't sure how to calculate the density of an object with multiple parts, examine the CD-ROM articles.)

Ballast disk	Density
A Copper	
B Aluminum	
C Brass	
D Titanium	
E Platinum	
F Zinc	

7. Examine the materials on the back counter of the lab. Use what you see to explain why the different ballast disks have different densities.

_____

_____

_____

_____

_____

_____

_____

_____

164 HOLT SCIENCE AND TECHNOLOGY INTERACTIVE EXPLORATIONS TEACHER'S GUIDE

Name _____ Date _____ Class _____

Exploration 2 Worksheet, continued

8. What is buoyant force? (If you're not sure, check out the CD-ROM articles.)

   _____

   _____

   _____

   _____

   _____

   _____

   _____

   _____

   _____

9. Describe the differences among underwater objects that are positively, negatively, and neutrally buoyant. (Hint: Check out the CD-ROM articles.)

   _____

   _____

   _____

   _____

   _____

   _____

   _____

   _____

   _____

   _____

Record your answers in the fax to Ms. Sittie.

EXPLORATION 2 • SEA THE LIGHT   165

Name _____ Date _____ Class _____

Exploration 2
Fax Form

FAX

To: Ms. Diane Sittie (FAX 817-555-4459)

From:

Date:

Subject: Sea the Light

Please complete the following chart:

METAL	MASS	VOLUME	DENSITY
Aluminum			
Brass			
Copper			
Platinum			
Titanium			
Zinc			

What is the density of the waterproof lamp base?
_____

Please indicate your metal selection for the ballast disk here: _____

Why did you pick this metal?
_____
_____
_____
_____
_____

166 HOLT SCIENCE AND TECHNOLOGY INTERACTIVE EXPLORATIONS TEACHER'S GUIDE

Sea the Light

The following articles can also be found by clicking the computer in the CD-ROM laboratory for Exploration 2:

- *A Dense Discussion*
- *Oh, Buoy!*
- *Diver Down*

Exploration 2
CD-ROM Articles

A Dense Discussion

What Is Density?

Density is a measure of the amount of matter in a given amount of space. Density is also defined as mass per unit volume. Within a given volume, dense materials have a lot of mass, whereas less-dense materials have less mass. For example, 5 mL of steel, a fairly dense material, has more mass than does 5 mL of cork, a less-dense material.

Because density is a physical property of a substance, objects can be distinguished by their densities. One way to distinguish between liquids, for example, is by a technique called layering. A liquid that is denser than another will sink below the less-dense liquid. For example, corn oil floats on top of water because corn oil is less dense than water.

Calculating Density

To determine the density of an object, you must first measure its mass and its volume. A balance can be used to measure the object's mass. A graduated cylinder can be used to measure the object's volume. The object's density can then be calculated using this simple equation: density = mass ÷ volume.

Suppose that you have a sample of lead that has a mass of 34 g and a volume of 3 mL. Inserting these measured values into the equation, you get the following:
density = 34 g ÷ 3 mL = 11.35 g/mL.

How can you determine the overall density of an object that consists of two parts with different densities? One way to do this would be to measure the mass and volume of each part and then use the following equation: total density = (mass of first part + mass of second part) ÷ (volume of first part + volume of second part).

Density Table

Here are density values for some common materials, including solids, liquids, and gases.

Material	Density (g/mL)
Helium (gas)	17.85×10^{-5}
Oxygen (gas)	13.31×10^{-4}
Ethyl alcohol (liquid)	0.79
Water (liquid)	1.00
Iron (solid)	7.87
Silver (solid)	10.50
Lead (solid)	11.34
Mercury (liquid)	13.55
Gold (solid)	19.31

Particles and Density

Different substances have different densities. This is partly due to the different masses of the particles that make up the substances. Imagine that you have several solid lead balls and several

EXPLORATION 2 • SEA THE LIGHT **167**

solid rubber balls, all of the same size. Now suppose that you fill one box with the lead balls and another box with the rubber balls. The balls in each box are packed identically and are the same size. As you can probably imagine, the box full of lead balls is heavier. Therefore, you can conclude that the box filled with lead balls is more dense than the box filled with rubber balls.

Density also depends on the structure of a material. A substance that has a great deal of space between its particles will be less dense than a substance with tightly packed particles. For example, graphite and diamond are both made of carbon. The carbon particles in each substance have the same chemical makeup and are the same size. However, the carbon particles in diamond are packed together much more tightly than they are in graphite. As a result, diamond is more dense than graphite.

Conditions That Influence Density

Certain conditions can influence the density of a substance. For example, a change in temperature can change the density of a substance. As a substance's temperature increases, the substance usually expands, or takes up more space, filling a larger volume. As a result, the density of the substance decreases because the same mass occupies a larger volume. On the other hand, a substance usually contracts as its temperature decreases. As a result, its density increases because the same mass occupies a smaller volume. Water is an exception to the effect of temperature on density. When water freezes, it expands rather than contracts. But we will talk about this later.

Another condition that affects the density of a substance is pressure. For example, if you squeeze a sponge, the particles of the sponge squeeze closer together, taking up less volume. This makes the density of the sponge increase. If you let go of the sponge, the pressure is removed and the particles return to their original placement, increasing volume and decreasing density.

A Sea of Different Densities

The composition of sea water varies around the world. Some regions of the ocean have a relatively high salt content; other regions have a relatively low salt content. At a given temperature and pressure, sea water with a high salt content is more dense than sea water with a low salt content.

Temperature and pressure also affect the density of sea water. Cold water, such as that found near the ocean floor and in the polar regions, is generally more dense than warm surface water or water found in tropical areas. The density of sea water also increases with depth because pressure increases with depth.

Oh, Buoy!

A Force in Fluids

Why do some things float while others sink? Well, fluids exert an upward force, called a **buoyant force,** on objects that are partially or completely submerged in them. If you have ever floated on an air mattress in a swimming pool, you've experienced this buoyant force. You can also feel it when you lift a heavy object, such as a brick, underwater. The brick seems a lot lighter underwater than it would above water. That's because the water exerts an upward buoyant force on the brick that makes the brick seem like it weighs less.

Buoyant Force (Part 1)

Buoyant force is due to the difference in the amount of fluid pressure between the top and the bottom of this column of water.

Perhaps you are wondering where buoyant force comes from. Look at the picture of a column of water in a lake. The bottom of the column receives greater pressure than does the top of the column because the bottom is supporting a larger amount of water above it.

The difference in the amount of force due to pressure on an object is known as the **buoyant force.** You may wonder why the buoyant force is always in the "up" direction. The buoyant force is the result of a difference in pressure. If an object is squeezed equally in every direction, it will not feel a force pushing it up, down, or sideways. Because the pressure in a fluid increases with depth, the bottom of an object in a fluid always receives more pressure than the top of the object. This is not the case for the sides of an object. When there is pressure on one side of an object, then there is equal pressure (at the same depth) on the other side of the object. Therefore, the only difference in pressure is from top to bottom. Buoyant force is an upward force because the pressure at the greater depth is always greater than the pressure at the lesser depth.

Buoyant Force (Part 2)

Now let's suppose that the underwater column is a tennis-ball canister. Again, the pressure at the bottom of the canister is greater than the pressure at the top of the canister, so there is an upward buoyant force. But now we have to consider something else: the tennis-ball canister displaces an amount of water equal to its own volume. Archimedes, a Greek scientist and mathematician, discovered that the buoyant force on an object is equal to the weight of the fluid displaced by the object. In this example, the buoyant force equals the weight of the water displaced by the tennis-ball canister.

The weight of the object itself does not determine the buoyant force. An object's weight comes into play only when the effect of the buoyant force on the object is considered. In other words, we consider the object's weight when we're trying to figure out whether an object will sink or float.

Buoyancy and Archimedes

An object immersed in a fluid experiences a buoyant force that is equal to the weight of the fluid displaced by the object. By this reasoning, an object will float if, and only if, it can displace a volume of liquid that is equal to its own weight.

Let's consider a steel ball and a steel boat, each with the same mass. Both are made of the same amount and type of material, but as you can probably guess, the boat will float and the ball will sink. Why? The difference is in the amount of water displaced by each object. The boat can displace a volume of water that weighs the same as the boat weighs. But because of its shape, the ball does not displace much water. Because the ball does not displace a volume of water that is equal to its own weight, it sinks.

Buoyancy and Density

An olive floats in the liquid in beaker *A* but sinks in the liquid in beaker *B*.

What would happen if you put identical objects in different liquids? Take a look at the olives in the beakers shown here.

In this example, each olive has the same volume and the same mass. So why does one olive float while the other one sinks? The answer has to do with the liquids in which they are placed. One liquid is more dense than the other. Can you figure out which one?

Remember that the buoyant force on an object is equal to the weight of the fluid it displaces. This being the case, the liquid in beaker *A* must be more dense than the liquid in beaker *B*. Why? If each olive displaces the same volume of liquid, the liquid in beaker *A* must weigh more. The only way it can weigh more is if it is more dense.

The Density of Water

Usually, solids are more dense than liquids because the particles in solids are closer together. So why does ice float in water?

The way in which individual water molecules are arranged affects the water's density. In a liquid state, the molecules are relatively close together. In a solid or frozen state, the molecules are locked into positions that are farther apart than the molecules are in the liquid state. As a result, the volume of the water increases when it freezes while the mass stays the same. That's why ice floats in water.

Controlling Buoyancy

How does a submarine control whether it sinks, floats, or stays in the same position underwater? The trick is to control the density of the submarine.

Built into the submarine are tanks that can be filled with water to give the submarine ballast. **Ballast** is a term used to describe anything heavy that is carried in ships or other structures to add mass. To make the submarine submerge, the tanks are filled with sea water. This addition of mass (without a change in the volume of the submarine) makes the submarine more dense. As a result, it sinks and can be described as **negatively buoyant.** When the submarine sinks to the depth at which its density equals the density of the surrounding water, it becomes **neutrally buoyant.** An object is neutrally buoyant when it neither sinks nor rises underwater. To bring the submarine to the water's surface, the water is pumped out of the tanks (and the tanks are filled with air), decreasing the overall density of the submarine. When its density is less than the density of the surrounding sea water, the submarine rises and can be described as **positively buoyant.**

Diver Down

A Brief History of Underwater Diving

For over 2,000 years, people have attempted to design mechanical means to stay underwater for as long as possible. Take a look at the paragraphs below for some momentous events in the history of underwater diving.

332 B.C.—Aristotle, the ancient Greek philosopher, describes a diving bell used by his prize student, Alexander the Great.

1500s—Renaissance artist and inventor Leonardo da Vinci designs a single diving system that combines air supply and buoyancy control.

1808—Friedrich von Drieberg invents the Triton apparatus, which features a backpack air reservoir supplied with air from above water. By nodding back and forth, the diver receives air through a valve.

1825—William James designs a diving system of closed tanks with compressed air. (The word *scuba* is an acronym for self contained underwater breathing apparatus.)

1911—The Davis False Lung, invented by Sir Robert Davis, saves lives worldwide in emergency submarine rescues. The device is a self-contained "rebreather" that provides crew members with enough air to swim to the water's surface.

1937—At the Paris International Exposition, divers demonstrate a scuba system that combines compressed air tanks with a valve that lets them regulate the amount of air they take in.

1943—Jacques Cousteau and Emile Gagnan invent the aqualung, the first safe and simple underwater breathing device.

Spotlight on Scuba Diving

In the 1940s and 1950s, Jacques Cousteau and his friends helped popularize scuba diving all over the world. His underwater films have provided a glimpse of the world that exists below the surface of the water.

In the United States, scuba-diving schools can teach you to use diving equipment safely. Minimum age requirements vary by scuba school and by course, but most require that you be at least 12 years old to become a junior certified diver. Open Water Diver is the most common certification for beginning scuba divers. It usually requires a minimum of 12 classroom hours, 12 hours of confined water training (in a swimming pool), and four to five open-water or ocean dives. Open Water Diver certification is a prerequisite for more-advanced diving courses, including Advanced Scuba Diver and Divemaster courses. Advanced scuba students can learn underwater photography, rescue diving, deep diving, cave diving, and search-and-recovery diving.

Exploration 3
Teacher's Notes

Stranger Than Friction

Key Concepts	Frictional force is the force that opposes motion when two objects are touching. The force of friction is not dependent on surface area.
Summary	Mr. Norm N. Cline has designed a new ride for his amusement park. The ride consists of a slide and several toboggans. He wants to know what material he should use to construct the slide and the bottom of the toboggans so that the amount of friction between the two will ensure a ride that is both exciting and safe. He also wants to know what size to make the toboggans.
Mission	Help the owner of an amusement park choose the best materials for a park ride.
Solution	Constructing the slide and the bottom of the toboggans out of stainless steel results in an amusement park ride that is both exciting and safe. The size of the toboggan has no effect on the slide's performance because frictional force is determined only by the normal force and the coefficient of friction between two surfaces, not by surface area.
Background	Many roller coasters have only one motorized mechanism—the conveyor belt that brings the train of cars to the top of the first hill. From there, only the force of gravity and the frictional force between the track and the wheels of the cars determine how fast the cars go down the hills, through the loops, and along the track. The taller a roller coaster track is at its initial height, the faster the velocity that the cars can achieve as they are accelerated by the force of gravity on their way down the first hill. Engineers continue to improve on the design of roller coasters to make them even more thrilling, more extreme, and faster. Some rides can send passengers zooming around at over 120 km/hr!
	There is no such thing as a 100-percent-efficient machine, and that certainly influences how many roller coasters are designed. At the crest of the first hill, which is the tallest point on many roller coasters, the potential energy of the coaster cars is greater than it will be at any other point during the entire ride. As the cars head down the first hill, the potential energy is converted into kinetic, heat, and sound energy. The total amount of energy is conserved, but only the kinetic energy can be converted back into potential energy. This kinetic energy is enough to carry the cars up another hill. However, because the amount of the cars' kinetic energy is less than the original amount of potential energy, the second hill cannot be as high as the first hill.

EXPLORATION 3 • STRANGER THAN FRICTION

Exploration 3 Teacher's Notes, continued

Teaching Strategies

The main goal of this Exploration is to help students understand what causes frictional force and how frictional force affects moving objects. To ensure that students understand how the coefficient of friction functions in determining frictional force, discuss as a class the information in the CD-ROM articles related to the mathematical expression for the coefficient of friction. Determining frictional force mathematically, the coefficient of friction (μ) is the multiplier value for the normal force (F_n) because $F_f = \mu F_n$. This may help students understand how smaller coefficients of friction result in smaller frictional forces for a given normal force. For example, a coefficient of friction of 0.70 and a normal force of 10 N produce a frictional force of 7 N ($F_f = 0.70 \times 10 \text{ N} = 7 \text{ N}$). If the coefficient of friction is 0.30 and the normal force remains 10 N, then the frictional force is 3 N ($F_f = 0.30 \times 10 \text{ N} = 3 \text{ N}$).

Another concept students must understand is that the surface area of materials in contact with one another has no bearing on frictional force. By exploring the equipment on the back counter in the lab, students should be able to grasp this concept and then apply it to what the different-sized toboggans demonstrate. Once students understand that surface area does not affect frictional force, they should realize that conducting experiments with every possible combination of materials and toboggan size is not necessary.

Bibliography for Teachers

"Coast Coaster." *U.S. News & World Report,* 120 (23): June 10, 1996, p. 21.

Wade, Bob. "Hot Wheels in the Laboratory." *The Physics Teacher,* 34 (3): March 1996, p. 150.

Young, Janet. "Complex Creations from Simple Machines." *Science Teacher,* 61 (1): January 1994, pp. 16–19.

Bibliography for Students

Koehl, Carla, and Sarah Van Boven. "Better and Faster Ways to Lose Your Lunch." *Newsweek,* 127 (22): May 27, 1996, p. 8.

Timney, Mark C. "Ups and Downs of Coaster Physics." *Boys' Life,* 86 (6): June 1996, p. 50.

Other Media

Energy at Work
Video or videodisc
Churchill Media
6677 N. Northwest Highway
Chicago, IL 60631
800-829-1900

In addition to the above video and videodisc, students may find relevant information about work and energy by exploring the Internet. Interested students can search for physics articles using keywords such as *work, energy,* and *machine efficiency.* Students can also access information about *roller coasters* by exploring the Internet.

Name _____ Date _____ Class _____

Exploration 3
Worksheet

Stranger Than Friction

1. Mr. Cline is hard at work on his design for a new amusement park ride. What information is he seeking from you?

   _____
   _____
   _____
   _____

2. Describe the equipment Dr. Labcoat has set up on the front lab table and the back counter.

   _____
   _____
   _____
   _____

3. What is frictional force, and how does it affect a moving object? (Hint: If you're not sure, check out the CD-ROM articles.)

   _____
   _____
   _____
   _____

4. What does the wall chart in the lab show about normal force and the coefficient of friction?

   _____
   _____
   _____
   _____
   _____
   _____

Name _____ Date _____ Class _____

Exploration 3 Worksheet, continued

5. Use the force meter on the back lab counter to find the force required to pull each block. Record your results below.

   _____
   _____
   _____

6. Does the amount of surface area touching the block affect the force required to pull it? Why or why not?

   _____
   _____
   _____
   _____

7. Conduct the necessary tests with the prototype for the Camelback Super Slide, and record your results in the table below. (Hint: It may not be necessary to try every possible combination.)

Material component for slide	Material component for toboggan	Toboggan size (cm)	Observations

Record your answers in the fax to Mr. Cline.

Name _____ Date _____ Class _____

Exploration 3
Fax Form

FAX

To: Mr. Norm N. Cline (FAX 281-555-5276)

From:

Date:

Subject: Stranger Than Friction

What material do you recommend for the construction of the slide?
_____

What material do you recommend for the construction of the toboggan?
_____

What is your recommendation regarding toboggan size?

100 cm
120 cm
140 cm
any of the above

What effect does the size of the toboggan have on the performance of the Camelback Super Slide? Explain.
_____
_____
_____
_____
_____
_____

EXPLORATION 3 • STRANGER THAN FRICTION

Stranger Than Friction

The following articles can also be found by clicking the computer in the CD-ROM laboratory for Exploration 3:

- *The Nature of Friction*
- *Friction by the Numbers*
- *That's Amusing!*

The Nature of Friction

Friction Fundamentals

Imagine you are pushing a box loaded with books across the floor. As you push, the floor seems to oppose the motion of the box. This opposition of motion is due to the friction between the box and the floor. **Frictional force** is the force that opposes motion between two surfaces that are touching. In order to slide the box of books across the floor, you have to overcome friction. When the force of your push on the moving box matches the force of friction, the box moves at a constant velocity; that is, the box is not accelerating due to unbalanced forces. The box continues to move at this velocity until you change the amount of force applied. If you stop pushing the box, the box stops because frictional force is no longer opposed.

The amount of friction between two surfaces depends on two factors: the type of surfaces that are in contact and how hard the surfaces are pushed together. The more books you load into the box, the more difficult it is to push the box across the floor. Likewise, if the floor is covered with rough carpet instead of with smooth wood, the box is harder to push. Frictional force does not usually depend on the size of the surfaces in contact or on how fast the surfaces are moving.

Different Kinds of Friction

Have you ever noticed that it is harder to start an object in motion than it is to keep it in motion? Think about sliding down a slide. At first you seem to "stick" to the slide, so you have to apply a large force to get started. This is because initially you must overcome the frictional force caused by static friction. **Static friction,** or starting friction, is the friction between objects that are in contact but not yet in motion. Once you overcome static friction, you are able to slide down easily. This is because once you start moving, the larger static frictional force gives way to the lesser kinetic frictional force. **Kinetic friction** is the friction between moving objects that are in contact. The kinetic frictional force is almost always less than the static frictional force.

The amount of kinetic friction between objects varies, depending on how the objects are touching. For example, sliding a box across a floor is an example of overcoming sliding kinetic frictional force. Putting wheels under the box would make the box easier to push. This is because you have changed the nature of the friction from sliding kinetic friction to rolling kinetic friction. Of course, the friction between the floor and the box isn't the only kind of friction that comes into play. Friction also exists between the box and the air. This type of friction is called fluid kinetic friction.

Decreasing and Increasing Friction

Sometimes it is desirable to decrease the amount of friction between two surfaces. For example, adding oil, grease, or other lubricants to a car engine is a way to reduce friction between the engine's moving parts. Reducing friction in a car engine protects the engine and keeps it from getting too hot. Another example of reducing friction is waxing downhill skis. The wax allows the skis to slide along the snow with less resistance, increasing the skier's maximum acceleration down a slope.

Other times, it is desirable to increase friction between surfaces. This is particularly useful when you want to stop an object like a car or a bicycle. The brakes on both of these vehicles create friction between brake pads and the turning wheels. When you pull on the brake lever of a bicycle, you are pressing rubber brake pads against the rim of the wheel. As a result, the wheel slows down.

Friction and Surface Area

How fast the brake pads stop the bicycle described in the previous section certainly depends on how hard you squeeze the brake lever. The stopping force also depends on the material from which the brake pads are made. Fresh new brake pads would provide a greater frictional force than old, hard brake pads. So how about larger brake pads? Will larger brake pads provide more frictional force than smaller ones? Believe it or not, the answer is no. Friction is not affected by the area of contact between surfaces.

Machines and Friction

Frictional force affects the operation of every machine. In fact, friction is necessary for many machines and mechanical systems to work at all. For example, the friction between a nail (a wedge) and the surface into which it is hammered, such as wood, keeps the nail in a sturdy position. Frictional force can also work against a machine. For example, kinetic frictional force opposes the motion of a box moving up an inclined plane.

Frictional forces affect the efficiency of machines. Some of the energy put into a machine or mechanical system is converted into heat and sound energy by the friction of moving parts. That means that large frictional forces can decrease the amount of energy available to do work, whereas smaller frictional forces can result in more energy being available to do work. In other words, by reducing friction, a machine's efficiency can be increased.

Friction by the Numbers

Important Forces

For a closer look at frictional force, consider a picture representation of what goes on when a box moves across a surface.

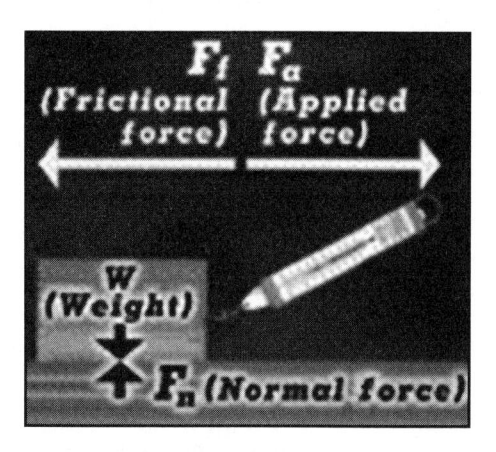

One obvious force involved in this situation is the weight of the box. Weight is a downward force, but the box is not accelerating downward, so there must be another force that opposes weight. This force is called the **normal force (F_n).** It is the perpendicular force pressing the two objects together. The greater the normal force on an object, the greater the frictional force. Because this box is on a horizontal surface, the normal force is equal in magnitude to the weight of the box. Notice that the arrows representing the applied force and the frictional force point in opposite directions. That's because frictional force opposes the motion of an object.

Coefficients of Friction

Remember that the types of surfaces in contact are very important in determining the amount of frictional force on an object. The friction between two surfaces can be expressed as a ratio. This ratio is called the coefficient of friction. The **coefficient of friction (μ)** is the ratio between the frictional force and the normal force. The amount of friction between two objects depends on whether an object is moving or starting in motion. Remember that it is more difficult to start an object in motion than it is to keep a moving object moving. We can express the difference in the amount of frictional force by using a coefficient of static friction and a coefficient of kinetic friction. The chart on the next page lists values for the coefficients of friction for various surfaces.

Exploration 3 CD-ROM articles, continued

Coefficients of Friction

Surfaces	Coefficient of static friction*	Coefficient of kinetic friction*
glass on glass	0.94	0.40
rubber tire on dry road	0.90	0.70
rubber tire on wet road	0.70	0.50
steel on ice	0.02	0.01
steel on steel	0.74	0.57
steel on steel with lubricant	0.20	0.12
teflon on teflon	0.40	0.04
wood on wood	0.50	0.40

*The values shown are averages.

Comparing the Numbers

A high coefficient of friction indicates a greater resistance to movement. A low coefficient of friction indicates a lower resistance to movement. In general, for a given pair of surfaces, the static coefficient of friction is greater than the kinetic coefficient of friction.

An Equation and Its Parts

Once you know the normal force and the coefficient of friction for a pair of objects, you can determine the frictional force. The equation for frictional force looks like this:

frictional force (F_f) = coefficient of friction (μ) × normal force (F_n)

Suppose that a wooden block is sitting on a horizontal wooden board and that you want to know how much force is necessary to get the block going. Obviously, you need to figure out what the starting frictional force is because that is the force that you will have to overcome.

To find the static frictional force, you would look up the coefficient of static friction for wood on wood, which is 0.50. Now you need to know the normal force. Because the block is horizontal, the normal force is equivalent to the block's weight. Suppose that value is 100 N. Now you have the values you need to solve the equation and find the force of static friction.

Given: $\mu = 0.50$
$F_n = 100$ N

Inserting these values into the equation $F_f = \mu \times F_n$, you have:

$$F_f = 0.50 \times 100 \text{ N} = 50 \text{ N}$$

Thus the force of static friction for the given wooden block on the given wooden surface is 50 N. That means you have to apply a force greater than 50 N in order to set the block in motion.

That's Amusing!

Record-Holding Roller Coasters

Engineers are constantly creating improved designs for rides that are guaranteed to take your breath away. Going through a loop, leaning through a tight turn, and plummeting from death-defying heights at top speeds all add to the excitement of riding roller coasters.

You can rank roller coasters any number of ways—from the fastest to the steepest to the tallest. But which coasters are the scariest? the most thrilling? Well, that is entirely up to you!

Steepest

The Ultra Twister, a steel roller coaster, at Astroworld, in Houston, Texas, is the steepest roller coaster in the world. It drops you at an angle of almost 90 degrees from horizontal. This ride is guaranteed to make you want to hold on to your hat!

Highest

At almost 80 m, the Megacoaster, at Fujiyama Park, in Japan, is the highest roller coaster in the world. It also has the longest drop of any roller coaster.

Fastest

The Steel Phantom, at Kennywood, in Pennsylvania, and the Desperado, at Buffalo Bill's Resort, in Nevada, are tied for the title of fastest roller coaster in the world. (They are also tied for highest roller coaster and longest drop in the United States.) These coasters travel at top speeds of about 128 km/hr. That's faster than the legal speed limit on most highways!

Exploration 4
Teacher's Notes

Latitude Attitude

Key Concepts	The Coriolis effect explains the apparent deflection of moving objects relative to the surface of the Earth. Airplane pilots must adjust their flight paths to account for the Coriolis effect.
Summary	Capt. Corey O. Lease has been commissioned to deliver supplies to Antarctica, where a team of scientists are studying the Mertz Glacier. Capt. Lease must make the trip immediately, but because the weather conditions there are terrible, she will not be able to land her plane at the site. Instead, she must drop the supplies as she flies over the area. Capt. Lease needs to know the best flight direction and average speed to use so that the supplies arrive safely at the site.
Mission	Advise Capt. Lease of the best flight plan to successfully complete her mission.
Solution	Flying at a compass direction of 230° and a speed of 725 km/hr is the best plan for Capt. Lease. Flying at a compass direction of 220° and a speed of 850 km/hr also allows Capt. Lease to deliver the supplies, but this combination uses too much fuel.
Background	The spherical shape of the Earth and the properties of centripetal acceleration contribute to the Coriolis effect. Centripetal acceleration is the acceleration needed to keep an object moving in a circle at a given radius. For a given centripetal acceleration, if the velocity of the spinning object increases, the radius of rotation increases as well ($a = v^2/r$). If the velocity of the object decreases, the radius decreases.
	The centripetal acceleration of objects on the Earth is caused by gravitational force. The radius of centripetal acceleration for the Earth is measured as a perpendicular line from the Earth's axis of rotation to any given point on the Earth's surface. As a result, there are different radii of centripetal acceleration for different points on the Earth's surface. For example, a point on the equator has a longer radius of centripetal acceleration than does a point near one of the poles. At a given radius, a point has a given velocity. For example, a point at 30°N latitude is rotating east at a velocity of 1,400 km/hr. Suppose an object at this point began moving east at 100 km/hr. The added vectors would give the object a resultant velocity of 1,500 km/hr. This faster velocity would tend to make the object increase its radius of motion. However, gravity keeps objects near the Earth's surface from flying out into space, and the only way the object can increase its radius of motion is to bend its path toward the equator (clockwise). This deflection is described as the Coriolis effect.

Exploration 4 Teacher's Notes, continued

Teaching Strategies

The concept of the Coriolis effect may be difficult for students to understand. You may need to perform various visual demonstrations with a globe to acquaint students with lines of latitude, the direction of the Earth's rotation, and the surface velocity of the Earth at different latitudes. As students work through the Exploration, remind them that Capt. Lease has a limited amount of fuel. Because fuel efficiency is important, students must recommend an appropriate average speed. Refer students to the CD-ROM articles to find out how speed and fuel are related.

As an extension of this Exploration, you may want to have students solve a similar situation that takes place in the Northern Hemisphere. Ask students how the Coriolis effect would influence a plane flying from the equator toward the North Pole. You may also want to explain why hurricanes rotate counterclockwise in the Northern Hemisphere and clockwise in the Southern Hemisphere, contrary to the behavior of other moving objects. Refer students to the CD-ROM articles to research this apparent contradiction.

Bibliography for Teachers

de Grasse Tyson, Neil. "The Coriolis Force." *Natural History,* 104 (3): March 1995, p. 76.

Lanken Dane, "Funnel Fury." *Canadian Geographic,* 116 (4): July/August 1996, p. 24.

Bibliography for Students

Gunther, Judith Anne. "Real-Time Weather." *Popular Science,* 246 (1): January 1995, p. 22.

Shepherdson, Nancy. "Controlling the Sky." *Boys' Life,* 85 (7): July 1995, p. 40.

Sweetman, Bill. "The Plane That Learns." *Popular Science,* 249 (1): July 1996, p. 40.

Other Media

Microsoft Flight Simulator 5.1 (for Macintosh and Windows 95)
CD-ROM or disk
Educational Resources
1550 Executive Dr.
Elgin, IL 60123
800-624-2926

Note: The above address is one of many locations from which you may order Microsoft products by mail. Microsoft products are also available at many retail stores. To locate the store nearest you, visit Microsoft's Web page at http://www.microsoft.com or call Microsoft's sales division at 800-426-9400.

Encourage students to search for relevant information about the Coriolis effect by exploring the Internet. Possible keywords include the following: *Coriolis effect, weather patterns, ocean currents,* and *hurricanes.* Students can also access information about *flying airplanes* and *flight paths* on the Internet.

Name _____ Date _____ Class _____

Exploration 4
Worksheet

Latitude Attitude

1. Capt. Corey O. Lease needs to get supplies to a group of scientists on the Mertz Glacier. What has she asked you to do for her?

   _____
   _____
   _____

2. In relation to Melbourne, where is the Mertz Glacier located?

   _____

3. How does the speed of the Earth's surface at the equator compare with the speed of the Earth's surface at the poles? (If you're not sure, check out the CD-ROM articles.)

   _____
   _____
   _____
   _____

4. How will you use the Accu-Flight Simulator to help Capt. Lease?

   _____
   _____
   _____
   _____
   _____
   _____
   _____

Name _____ Date _____ Class _____

Exploration 4 Worksheet, continued

5. Use the Accu-Flight Simulator to determine the results of using different combinations of flight direction and average airspeed. Record your results in the table below.

Flight direction	Airspeed (km/hr)	Final location of the plane
130°	725	
140°	725	
150°	725	
180°	725	
210°	725	
220°	725	
230°	725	
130°	850	
140°	850	
150°	850	
180°	850	
210°	850	
220°	850	
230°	850	

6. What does the equipment on the lab's back counter demonstrate?

_____

_____

_____

_____

7. Describe how moving objects are affected by the Coriolis effect. (If you're not sure, check out the CD-ROM articles.)

_____

_____

_____

_____

_____

Record your answers in the fax to Capt. Lease.

EXPLORATION 4 • LATITUDE ATTITUDE   **183**

Name _____ Date _____ Class _____

Exploration 4
Fax Form

FAX

To: Capt. Corey O. Lease (FAX 011-612-555-7996)

From:

Date:

Subject: Latitude Attitude

In which compass direction should I aim my plane?

| 180° (due south) | 130° | 140° | 150° | 210° | 220° | 230° |

How fast should I fly the plane?

| 725 km/hr | 850 km/hr |

Please explain why it is necessary for me to fly the plane in the recommended direction in order to reach the Mertz Glacier.

_____
_____
_____
_____
_____
_____
_____
_____
_____
_____

184 HOLT SCIENCE AND TECHNOLOGY INTERACTIVE EXPLORATIONS TEACHER'S GUIDE

> **Exploration 4**
> CD-ROM Articles

> ### Latitude Attitude
> The following articles can also be found by clicking the computer in the CD-ROM laboratory for Exploration 4:
> - *The Coriolis Effect*
> - *Wind, Water, and Weather*
> - *What a Pilot Needs to Know*
> - *Antarctica*

The Coriolis Effect

A Little Latitude

Lines of latitude are imaginary parallel lines that run east to west around the Earth. The equator, at a latitude of 0 degrees, divides the Earth in two halves, called the Northern Hemisphere and the Southern Hemisphere.

What Is the Coriolis Effect?

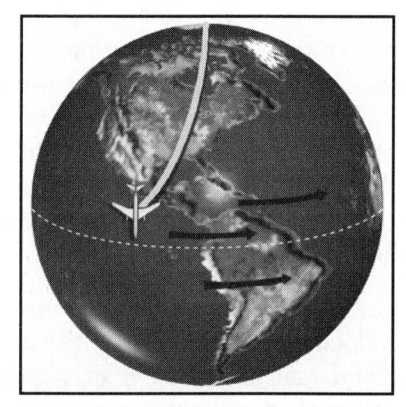

Because of the Coriolis effect, objects traveling south in the Northern Hemisphere appear to curve to the right (clockwise).

In 1835, Gustave-Gaspard de Coriolis, a French mathematician, explained that an object moving in a straight line perpendicular to the Earth's equator will appear to curve. This effect, called the **Coriolis effect,** is the apparent bending, or deflecting, of the path of a moving object due to the Earth's rotation. The Coriolis effect influences the paths of moving objects, such as airplanes, winds, and weather systems. In the Northern Hemisphere, objects moving north or south seem to veer to their right (clockwise). In the Southern Hemisphere, moving objects tend to veer to their left (counterclockwise).

So what causes this curving effect? The circumference of the Earth at the equator is larger than the circumference of the Earth at 30°, 60°, or at the poles. That means that a point on the equator travels farther during one of Earth's rotations (one day) than does a point at 30°, 60°, or near one of the poles. As a result, points on the Earth's surface near the equator move faster (about 1670 km/hr) than do points at 30° (about 1400 km/hr), 60° (about 800 km/hr), or at one of the poles (effectively 0 km/hr).

You can see how the Coriolis effect works by slowly spinning a record turntable clockwise. As the turntable rotates, imagine what would happen if you traced a line from the outside of the turntable to the center. You would get a line that curves to the left. Even though you trace a straight line, the clockwise rotation of the turntable causes the line that you trace to appear as a curve. Likewise, even though an object moving above the Earth's surface may follow a straight path, we will actually see a curve. That's the Coriolis effect.

Wind, Water, and Weather

Wind Patterns

Generally, air flows from areas of high pressure to areas of low pressure. Differences in pressure cause air to circulate between the poles and the equator, creating wind patterns. As the Earth rotates, the Coriolis effect influences the northerly or southerly path of the wind. In the Northern Hemisphere, winds circulate clockwise, while winds circulate counterclockwise in the Southern Hemisphere.

Ocean Currents

Winds that blow across the ocean's surface transfer energy to the ocean water, resulting in strong surface currents. For example, the trade winds create currents that affect the movement of ocean water near the equator. These ocean currents, as well as other ocean currents, are

EXPLORATION 4 • LATITUDE ATTITUDE   185

deflected by the Coriolis effect so that the currents flow clockwise in the Northern Hemisphere and counterclockwise in the Southern Hemisphere.

The Coriolis Effect and Weather Patterns

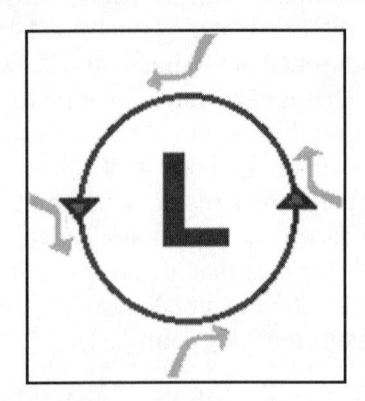

The movement of air masses causes most weather patterns, including thunderstorms, warm and cold fronts, and hurricanes. The Coriolis effect causes the deflection of air masses as air moves in the atmosphere. For example, in the Northern Hemisphere, parcels of high-pressure air moving toward a low-pressure system will be deflected to the right. This causes the low-pressure system to spin counterclockwise. For this reason, storm systems and hurricanes in the Northern Hemisphere spin counterclockwise, while in the Southern Hemisphere they spin clockwise.

What a Pilot Needs to Know

Speed and Fuel

One way to measure an airplane's speed is by ground speed. **Ground speed** is the speed of the plane with respect to the ground. Another measure is air speed. **Air speed** is the speed of the plane with respect to the air. If there is no wind (that is, if the air is not moving), then the air speed of the plane equals its ground speed. The air-speed indicator in the cockpit of a plane helps a pilot determine the plane's air speed.

The speed at which a plane flies affects the amount of fuel that the plane needs to reach its destination. Pilots calculate the amount of fuel necessary to reach a destination at ground speed, then they adjust these calculations to account for changes in speed due to winds.

Winds blowing in the direction that the plane is flying allow the pilot to fly farther with less fuel. On the other hand, winds that blow against the plane force a pilot to use more fuel to travel the same distance. Flying faster requires more power, and, therefore, more fuel.

Flight Path

An aeronautical chart is very important in air navigation.

When you plan a road trip, you can look at a road map to determine in what direction you need to drive to reach various destinations. However, if you were to follow those same directions in an airplane, you might not end up in the place you expected. Why? When you travel in the air (especially over great distances), you must account for the Coriolis effect.

When a plane sits motionless on the runway, it still has a velocity because the Earth's surface is rotating west to east at a given velocity. When the plane takes off, it still has that velocity with respect to the runway. As the plane begins to travel north or south, the plane will have a different velocity with respect to the Earth's surface because the surface velocity of the Earth is different at different latitudes. As a result, the plane will have a different velocity with respect to the Earth's surface. An airplane flying north from the equator will seem to move east because the surface velocity of the Earth decreases as latitude increases (toward the North Pole). If the pilot follows a straight path north, the plane will end up east of its destination. An airplane flying south from the North Pole will seem to move west because the surface velocity of the Earth increases as latitude decreases (toward the

186   HOLT SCIENCE AND TECHNOLOGY INTERACTIVE EXPLORATIONS TEACHER'S GUIDE

equator). This apparent deflection is an example of the Coriolis effect. Pilots must compensate for the Coriolis effect when planning their flight paths.

Suppose an airplane leaves Quito, Ecuador, near the equator, and flies directly toward New Orleans, Louisiana, about 30°N latitude. The eastward velocity of the Earth's surface is approximately 1670 km/hr at Quito. As the plane heads toward New Orleans, it seems to have some eastward velocity because the Earth's surface does not rotate as fast near New Orleans as it does at Quito. So the plane is actually traveling a little bit east (from a perspective above the Earth's surface). This apparent change in velocity has an effect on where the plane ends up. If the plane maintained its initial flight direction, it would end up landing east of New Orleans. To compensate for this, a pilot would aim the plane slightly northwest to ensure that the plane reaches its destination of New Orleans.

So what does this mean in terms of planning a flight direction and flight path? It means that it is necessary to adjust for the Coriolis effect when planning the direction of flight. Luckily, pilots can use computers to calculate the necessary adjustments when creating flight paths.

Antarctica

The Frozen Continent

The continent of Antarctica, an area that measures approximately 13 million square km, is almost entirely covered by a thick sheet of ice. The ice sheet was created by the buildup of snow over millions of years, and it contains over 70 percent of the world's freshwater supply. As the ice covering Antarctica piles up, it turns into glaciers and ice rivers that flow to the sea. A vast current called the Antarctic Circumpolar Current moves the cold waters into the rest of the world's oceans.

Antarctica is surrounded by the Pacific, Atlantic, and Indian Oceans. The Antarctic landmass is divided by the Transantarctic Mountains, and its geological features include an active volcano named Mount Erebus and the geographic and magnetic south poles. The Mertz and Innis Glaciers extend into eastern Antarctica just north of the south magnetic pole.

Who Wants to Live in Antarctica?

The frozen land and frigid weather mean little life thrives on Antarctica.

The South Pole region has an annual mean temperature of −49°C, and Antarctica holds the world record for cold temperature, −89.2°C. Winter lasts from mid-March to mid-September. During this time, the continent is submerged in darkness. So who on Earth would want to live in Antarctica?

Despite the extreme cold, scientists from all over the world gather regularly at a number of Antarctic research stations. At the South Pole Station, for example, astronomers study galaxy and star formation while astrophysicists keep an eye on the ozone layer. The McMurdo Dry Valleys are another hot spot for research on Antarctica. This unusual expanse of ice-free land, with its mountain ranges, meltwater streams, and arid terrain, draws botanists, geochemists, biologists, ecologists, and other scientists.

Exploration 5
Teacher's Notes

Tunnel Vision

Key Concepts	Light bulbs connected in a parallel circuit are brighter than the same bulbs connected in a series circuit. Too much current flowing through a circuit can result in a burnout of the output device(s).
Summary	Seymore Rhodes, an aspiring inventor and bicycling enthusiast, is working on an electrical circuit for his bicycle helmet that will make his ride to school a safer one. Seymore has to ride his mountain bike through a dark tunnel on the way to school, and he is worried that drivers may not see him while he is riding through the tunnel. He has designed a circuit with three red lights that can attach to his bike helmet. However, the lights are too dim. Seymore needs to know how to make the lights brighter while keeping the circuit lightweight.
Mission	Advise a young inventor about the best circuit to use on his bicycle helmet.
Solution	Connecting the three bulbs in a parallel circuit powered by one battery is the best combination for Seymore's circuit. With this combination, the bulbs are very bright and the circuit is still lightweight.
Background	A typical cable used to wire houses for electricity is composed of three wires enclosed in a plastic casing. Two of the wires, one covered with gray insulation and one covered with blue insulation, carry alternating current to and from the outlets. The third wire is not insulated and acts as a safety feature. It is usually connected to a copper bar driven into the ground outside the house. If there is leakage of current, the electricity flows to the ground, preventing electrocution. Average homes are wired with four, five, or more parallel circuits. Several outlets are connected to each circuit. If too many appliances on one circuit are turned on at the same time, too much electric current moves through the wire. This can cause an overload, and that circuit will shut down, cutting off the electricity.
	Fuses are designed to protect against overloads. A fuse is a safety device containing a short strip of metal with a low melting point. If too much current passes through, the metal melts, causing the fuse to "blow." This causes a break in the circuit, and the current no longer flows. Another device that protects circuits from overloads is a circuit breaker. One type of circuit breaker is a switch attached to a bimetallic strip. When the metal gets hot, it bends. This opens, or "trips," the circuit. This action does not harm the circuit breaker. After the overload has been corrected, the circuit breaker can be reset.

Exploration 5 Teacher's Notes, continued

Teaching Strategies

Students may need extra time to become familiar with the circuit board in this Exploration. Explain to them that they will create circuits by closing switches, not by arranging wires. You may want to demonstrate how to create several different circuits. Students may also benefit from drawing circuit diagrams so that they can determine what types of circuits they create. In addition, students can create numerous circuits that contain only one or two light bulbs, but these will not qualify as possibilities for Seymore's circuit. In order to increase students' efficiency, you may want to suggest at the outset that students focus on creating circuits using all three light bulbs. As students begin to create circuits in the lab, remind them to note not only whether the bulbs light up but also how bright they are.

As an extension of this Exploration, you may wish to create similar electric circuits in your classroom. This may help students visualize what they are doing in Dr. Labcoat's lab. Using flashlight bulbs, batteries, and wires with alligator clips, they can easily construct the different circuits from the Exploration. If students construct the circuits themselves, remind them to be careful about giving too much voltage to circuits. Too much voltage can burn out the bulbs.

Bibliography for Teachers

Lipkin, Richard. "Teeny-Weeny Transistors." *Science News,* 147 (18): May 6, 1995, p. 287.

Steiger, Walter R., and Suk R. Hwang. "A Series-Parallel Demonstration." *The Physics Teacher,* 33 (9): December 1995, p. 590.

Bibliography for Students

Humberstone, E. *Electricity and Magnetism.* Usborne/EDC, 1994.

Kempter, Joseph. "Power to Go." *Astronomy,* 24 (1): January 1996, pp. 86–87.

Reis, Ronald A. "Get Charged." *Boys' Life,* 86 (1): January 1996, p. 53.

Other Media

Learning All About Electricity and Magnetism
CD-ROM for MAC or MS-DOS
Queue, Inc.
338 Commerce Drive
Fairfield, CT 06432
203-335-0906
800-232-2224

In addition to the above CD-ROM, students may also find relevant information about electricity and magnetism by exploring the Internet. Interested students can search for articles using keywords such as the following: *electricity, magnetism, electric current, electric circuits,* and *Ohm's law.*

Name _____ Date _____ Class _____

Exploration 5
Worksheet

Tunnel Vision

1. Seymore Rhodes wants to make his mountain-bike ride to school a safer one. What advice does he need from you?

2. What are the necessary components of an electric circuit? (If you're not sure, check out the CD-ROM articles.)

3. Look at the diagram that Seymore drew of his circuit. What kind of circuit is it?

4. How is a series circuit different from a parallel circuit? (Hint: Check out the equipment on the lab's back counter.)

Name _____ Date _____ Class _____

Exploration 5 Worksheet, continued

5. Examine the equipment on the lab's front table. How will you use the switches and the batteries to help you answer Seymore's questions?

_____

_____

_____

_____

_____

6. Create all of the necessary circuits you need to answer Seymore's questions. In the table below, record the switches you close, the number of batteries you use, and the type of circuit you create. Use the fourth column to record your observations about the light bulbs.

Switches closed	Number of batteries	Type of circuit	Effect on light bulbs

Name _____ Date _____ Class _____

Exploration 5 Worksheet, continued

7. What effect does the number of batteries have on your circuits?

   _____
   _____
   _____
   _____
   _____
   _____
   _____
   _____
   _____
   _____

8. What happens when too much current flows through a circuit?

   _____
   _____
   _____
   _____
   _____
   _____

9. What is the relationship between current, voltage, and resistance? (If you're not sure, check out the CD-ROM articles.)

   _____
   _____
   _____
   _____
   _____
   _____
   _____
   _____

Record your answers in the fax to Seymore Rhodes.

Name _____ Date _____ Class _____

Exploration 5
Fax Form

FAX

To: Seymore Rhodes (FAX 406-555-7209)

From:

Date:

Subject: Tunnel Vision

How many batteries should I use? | 1 | 2 |

Please describe the best way to wire the lights for my helmet.

_____

_____

_____

_____

_____

_____

_____

_____

_____

✂︎·

For Internal Use Only

Please answer the following questions for my laboratory records. Scientists must always keep good records. *Dr. Crystal Labcoat*

What is the best type of circuit for Seymore's lights?

| SERIES | SERIES-PARALLEL | PARALLEL |

EXPLORATION 5 • TUNNEL VISION

Name _____ Date _____ Class _____

Exploration 5 Fax Form, continued

Describe the changes that you made to Seymore's original circuit.

> ## Tunnel Vision
>
> The following articles can also be found by clicking the computer in the CD-ROM laboratory for Exploration 5:
> - *The Essentials of Electricity*
> - *Circuits—Making the Connection*

Exploration 5
CD-ROM Articles

The Essentials of Electricity

What Is Electric Current?

A bolt of lightning can be described as a big electric spark in the sky. More specifically, it is an electric current passing from one cloud to another cloud or to the Earth, or from the Earth to a cloud. An **electric current** is the flow of electrons from an object that has many electrons to another object that has too few electrons. The uneven distribution of electrons in the sky is a result of friction between the moving clouds and the surrounding air. To understand this process, think about what happens when you scuff your feet on a carpet and reach for a doorknob. Ouch! The shock you experience is the flow of electrons, or electric current, from your hand to the doorknob.

Although these examples of electric current occur naturally, they are not useful in our everyday lives. To make electric current useful, an electric circuit is required. An **electric circuit** is a continuous pathway for an electric current. The components of a simple circuit include a source of electricity, such as a battery; an output device, such as a light bulb; and wires. Many circuits also include switches, which allow the circuit to be turned on and off. When the switch is off, the pathway of the circuit is interrupted. This situation is called an **open circuit,** and no current can flow. When the switch is turned on, the circuit is complete and the current flows. This situation is called a **closed circuit.**

How Does a Flashlight Work?

A flashlight is an example of a simple circuit. A flashlight generally consists of a plastic case, two batteries, a flashlight bulb, a switch, and a metal strip. When the flashlight is off, the switch is open, and the metal strip does not touch the flashlight bulb. When you turn the flashlight on, the switch closes the circuit by moving the metal strip so that it touches the flashlight bulb. Electrons then flow from the negative end of the battery, through the metal strip, through the filament in the bulb, and then to the positive end of the battery. As the current flows through the filament, the filament heats up and glows to produce light.

Circuits—Making the Connection

Circuit Symbols

Drawing a diagram is a useful way to describe a circuit. Remember that circuits can involve various output devices, such as light bulbs, toasters, hair dryers, microwave ovens, and almost anything else you can think of that you plug in. In the following sections, we will discuss various kinds of circuits, but we will use light bulbs in every example.

Series Circuits

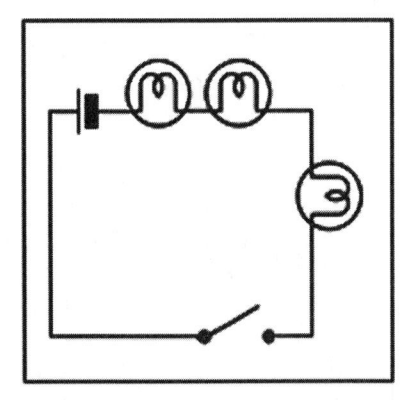

In a series circuit, there is only one pathway for the current to follow. If there is a break in the circuit, the flow of current is disrupted, and none of the bulbs light. The fact that the bulbs are connected in series also affects the brightness of the bulbs. Because the light bulbs share the same current, three bulbs in series are dimmer than one single bulb connected to the battery.

EXPLORATION 5 • TUNNEL VISION   **195**

Parallel Circuits

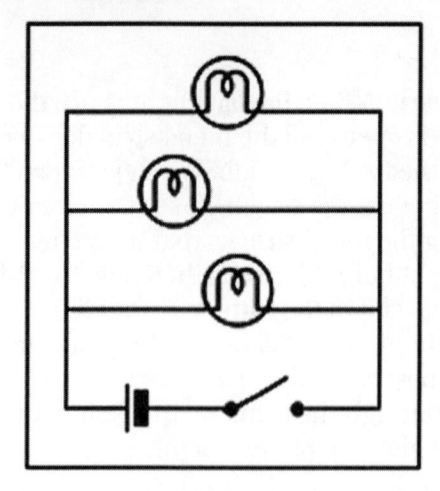

In a parallel circuit, current is divided into two or more branches. When current reaches a branching point, some current goes along one branch and some goes to another branch. If a break occurs in one branch, current can continue to flow through other branches and the bulbs along those branches stay lit. The fact that the bulbs are connected in parallel also determines the brightness of the bulbs. The battery powers each bulb individually, so each bulb emits its standard brightness. In other words, three bulbs in parallel are just as bright as one single bulb connected to the battery.

Series-Parallel Circuits

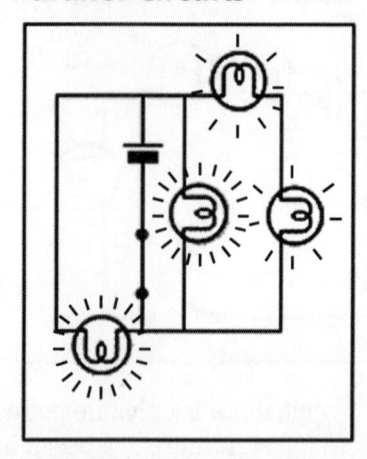

Series-parallel circuits contain elements of both series and parallel circuits.

Path of Least Resistance

Electric current tends to follow the path of least resistance. **Resistance** can be thought of as electrical friction—it is the opposition to current in a circuit. All output devices (such as light bulbs, etc.) offer some resistance to electric current. In a circuit that involves light bulbs, electric current tends to follow the path with the fewest bulbs or with bulbs with the least resistance. Keep in mind that some of the current flows through each complete pathway. How much current flows through a particular pathway depends on the resistance of that pathway as compared to the other pathways. In a series circuit, current follows only one pathway, so each bulb lights equally if each bulb provides the same resistance. In a parallel circuit, however, more current always goes into the branches of the circuit with lower resistance, and less current goes into the branches with higher resistance.

Resistance can be calculated using Ohm's law, which describes the relationship between voltage, current, and resistance in an electric circuit. Ohm's law states:

$$\text{current (I)} = \frac{\text{voltage (V)}}{\text{resistance (R)}}$$

$$\text{or } I = \frac{V}{R}$$

You can use the equation to solve for any one of the variables. To find the resistance in a circuit, you would divide V by I. The ratio V/I demonstrates that, for a given resistance, current and voltage are directly proportional. This allows you to predict the effect of changes in the circuit. Doubling the voltage in a circuit will double the current in the circuit.

Circuit Failure

Although circuits are easy to design and construct, sometimes problems arise that cause the circuit to fail or to not operate properly. For example, if too much current is introduced into a circuit involving a light bulb, the bulb can burn out. Remember that Ohm's law states that current and voltage are directly proportional. That means that for a circuit with a given resistance, an increase in voltage means an increase in current. In our example of the flashlight, that means that too much voltage from the batteries can burn out the bulb.

Another example of circuit failure is a short circuit. A **short circuit** occurs when the bulbs and wires are connected in such a way that the current bypasses the bulb. When the current takes such a short cut, the bulb won't light.

Exploration 6
Teacher's Notes

Sound Bite!

Key Concepts	Sound waves can be described by their wavelength, frequency, and phase. Active noise control uses deconstructive interference to reduce the amplitude of a sound wave.
Summary	Mr. Cy Lintz, the owner of a pet store, is losing business because his usually mild-mannered guinea pigs have been biting the customers! He suspects that the cause of this biting outbreak is the constant hum of freezers coming from the new ice cream shop next door. Mr. Lintz knows that guinea pigs become highly agitated when exposed to loud noises, and he believes this is why the guinea pigs are displaying this aggressive behavior. Mr. Lintz has read about a new technology used to eliminate noise pollution and would like to know how it might be applied to solve his problem.
Mission	Apply the concept of active noise control in an attempt to soothe some agitated guinea pigs.
Solution	A sound in phase 2 with a frequency of 225 Hz and a low amplitude will eliminate the offending sound. Its waves destructively interfere with those of the offending sound (which is in phase 1 and has a frequency of 225 Hz and a low amplitude).
Background	Your ear is specialized for picking up sound waves and transmitting them to your brain. A sound, such as a telephone ring, causes particles in the air around the phone to vibrate. These vibrations, or sound waves, travel through the air in longitudinal waves. The outer ear, or pinna, catches the sound waves and directs them into the ear canal. When the sound waves reach the eardrum at the end of the canal, the eardrum begins to vibrate. The vibrations are passed along to three bones, the hammer, anvil, and stirrup, on the other side of the eardrum. These bones act like levers, multiplying the force of the vibrations. When the stirrup vibrates, the energy of the vibrations passes through the oval window into the cochlea, where it causes waves in the fluid there. Along the length of the cochlea are tiny hair cells that vibrate in response to the frequency of the waves. Nerve cells at the base of the hair cells change the vibrations into electrical impulses. These impulses are carried along the auditory nerve to the brain, where they are interpreted as the ring of a telephone.

Exploration 6 Teacher's Notes, continued

Teaching Strategies

Although this Exploration requires students to analyze aspects of sound by examining transverse waves on oscilloscope screens, you may need to emphasize that sound waves are actually longitudinal waves. You may refer students to the wall chart and discuss how oscilloscopes convert longitudinal waves to transverse waves. Converting longitudinal waves to transverse waves on an oscilloscope screen allows you to "see" the frequency and amplitude of a sound. In addition, you may want to explain the function of the two phase buttons in the Exploration. They do not represent the only possible positions for a transverse wave. Refer students to the CD-ROM articles for a discussion of waves that are in phase or out of phase. Explain that the phase options in this Exploration represent an opportunity to put the waves exactly in phase or exactly out of phase.

As an extension of this Exploration, you may wish to conduct an investigative discussion of how sounds travel through different mediums. In this Exploration, the offending sound traveled through a connecting vent between the pet store and the ice cream shop. You could determine how the situation would be different if the humming noise were instead passing through an adjoining wall, the ceiling, or the floor. *(The sound could be harder to hear, more muffled, and more difficult to analyze in terms of its frequency and amplitude, for example.)* Ask students how such a change would affect how Mr. Lintz could implement active noise control. *(He couldn't simply generate a sound through the vent, for example.)*

Bibliography for Teachers

Sungar, Nilgun. "Teaching the Superposition of Waves." *The Physics Teacher,* 34 (3): April 1996, p. 236.

Wolkomir, Richard. "Decibel by Decibel, Reducing the Din to a Very Dull Roar." *Smithsonian,* 26 (11): February 1996, pp. 56–65.

Bibliography for Students

Foster, Edward J. "Switched On Silence." *Popular Science,* 245 (1): July 1994, p. 33.

Johnson, Julie. "Beating Bedlam." *New Scientist,* 144 (1947): October 15, 1994, pp. 44–47.

Other Media

Standing Waves and the Principle of Superposition
Video
Encyclopædia Britannica Educational Corporation
310 S. Michigan Ave.
Chicago, IL 60604-9839
800-554-9862

In addition to the above video, students may find relevant information about sound and sound waves by exploring the Internet. Interested students can search for articles using keywords such as *sound, sound waves, constructive interference, deconstructive interference,* and *acoustics.* Students can also access information about *active noise control* on the Internet.

Name _____ Date _____ Class _____

Exploration 6
Worksheet

Sound Bite!

1. Mr. Lintz is worried about the aggressive behavior of his guinea pigs. Describe his problem and what he has asked you to do.

2. Examine the equipment on the back counter of the lab, and then describe what sound waves are.

3. What purpose does an oscilloscope serve? (Hint: Check out the wall chart.)

4. What is destructive interference? (If you're not sure, check out the CD-ROM articles.)

EXPLORATION 6 • SOUND BITE! 199

Name _____ Date _____ Class _____

Exploration 6 Worksheet, continued

5. Conduct your experiment using the lab equipment. In the table below, record the frequency, amplitude, and phase you selected for the generated sound. Use the fourth column to describe the wave you see in the center oscilloscope screen when the generated sound is combined with the offending sound.

Frequency (Hz)	Amplitude	Phase	Results

Name _____ Date _____ Class _____

Exploration 6 Worksheet, continued

6. Based on your results, what are the offending sound's frequency, amplitude, and phase?

_____

_____

_____

_____

7. What does *phase* refer to? (If you're not sure, check out the CD-ROM articles.)

_____

_____

_____

_____

_____

Record your answers in the fax to Mr. Lintz.

Name _____ Date _____ Class _____

Exploration 6
Fax Form

FAX

To: Mr. Cy Lintz (FAX 707-555-8988)

From:

Date:

Subject: Sound Bite!

Which settings eliminate Mr. Lintz's noise pollution?

Frequency (hertz)	175	225	275	325
Amplitude	low	medium	high	
Phase	1	2		

For Internal Use Only

Please answer the following questions for my laboratory records. Scientists must always keep good records *Dr. Crystal Labcoat*

Explain how sound travels.

_____
_____
_____
_____
_____
_____
_____
_____
_____
_____

202 HOLT SCIENCE AND TECHNOLOGY INTERACTIVE EXPLORATIONS TEACHER'S GUIDE

Name _____ Date _____ Class _____

Exploration 6 Fax Form, continued

Explain how active noise control is used to eliminate noise pollution.

> **Sound Bite!**
>
> The following articles can also be found by clicking the computer in the CD-ROM laboratory for Exploration 6:
> - *Makin' Waves*
> - *Wave Basics*
> - *Sound Advice*

Exploration 6
CD-ROM Articles

Makin' Waves

Good Vibrations

All sound is produced by vibrations. In the case of a bumblebee, you can see the vibrations in the movement of its wings. You can also see the vibrations of a guitar string that has been plucked. Most sound, however, is produced by vibrations that are too small and too fast to see with the unaided eye.

Sound begins when a vibrating object pushes on the particles around it. The particles near the object become more dense (more tightly packed) than the other particles around them. The dense particles then spread apart, pushing against other particles. This causes the other particles to crowd together and then spread apart in a back-and-forth motion. Particles continue to crowd together and spread apart in succession as the sound travels away from its source. The result is waves of sound.

Sound can travel through different mediums. Usually, the sounds that we hear are caused by sound waves that travel through the air, but we can also hear sound that travels through liquids and solids. Have you ever listened to sounds underwater in a swimming pool? While underwater, you can hear the sounds of shouts and splashes as sound waves pass through the water. Sound can also pass through solid materials. For example, you can hear voices through a wooden door. It is impossible for sound to travel where there are no particles to vibrate, so sound cannot travel through space or through a vacuum.

Transverse Waves

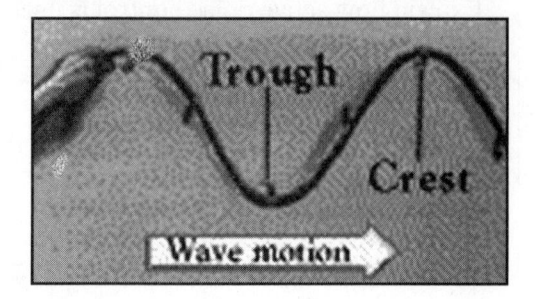

A **wave** is a disturbance that transmits energy as it travels through matter or space. A **transverse wave** is a wave in which the motion of the particles of the medium is perpendicular to the path of the wave. A transverse wave has a **crest,** the uppermost part of the wave, and a **trough,** the lowest part of the wave. You can generate a transverse wave by holding the end of a rope in your hand and moving your hand up and down. The energy moves along the wave, but the medium itself does not move. In other words, as you move the rope up and down, the rope itself does not travel horizontally, but you do send energy along the length of the rope.

Longitudinal Waves

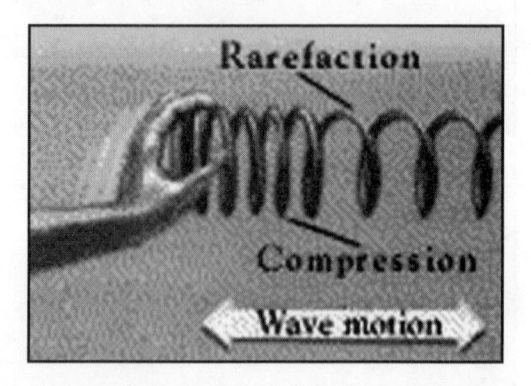

Not all waves move up and down like transverse waves. **Longitudinal waves** move back and forth, parallel to the direction of motion of the wave. Unlike transverse waves, longitudinal waves do not have crests or troughs. Instead, they have regions of **compression,** in which particles are pushed close together, and regions of **rarefaction,** in which particles are pulled apart from each

204 HOLT SCIENCE AND TECHNOLOGY INTERACTIVE EXPLORATIONS TEACHER'S GUIDE

other. Longitudinal waves consist of alternating compressions and rarefactions moving through a medium. If you alternately compress and stretch a spring, you can observe longitudinal waves. Notice again that the particles of the spring are not carried along with the compressions and rarefactions of the wave; they pass the energy of the wave along to the neighboring particles and return to their original position.

Wave Basics

Amplitude

Amplitude refers to the height of the wave. For a transverse wave, the amplitude is half the vertical distance between a crest and a trough. For a longitudinal wave, the amplitude is represented by the relative amount of compression and rarefaction the wave experiences. For both transverse and longitudinal waves, the greater the energy that the wave carries, the greater the wave's amplitude.

Wavelength

Wavelength refers to the distance between two identical points on neighboring waves. You can measure wavelength by measuring the distance between two crests or two troughs of a transverse wave or by measuring the distance between two compressions or two rarefactions of a longitudinal wave.

Frequency

Frequency refers to the number of waves produced in a given time. Frequency is usually measured in Hertz (Hz), the number of vibrations per second. Something that vibrates at a rate of 100 Hz vibrates 100 times per second. You could measure frequency in a transverse wave by counting the number of crests or troughs that pass a point in a given amount of time. For a longitudinal wave, you could count the number of compressions or rarefactions that go by a certain point in a given amount of time.

Wave Speed

Wave speed refers to the speed at which a wave travels and is usually measured in meters per second. The speed of a wave can be measured by observing how fast a single compression or crest moves through a medium. The speed of sound in air is about 346 m/s. In general, sound waves travel even faster through liquids and solids because liquids and solids have less space between particles.

Phase

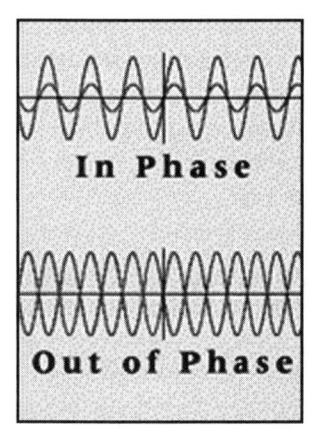

Phase is a term used to describe the positions of the crests and troughs of two individual waves relative to each other. For example, two waves of a given frequency are said to be in phase if the crests and troughs of one wave match up with the crests and troughs of the other wave. Two waves of a given frequency are said to be out of phase when the crests and troughs of the two waves do not match up.

Wave Interference

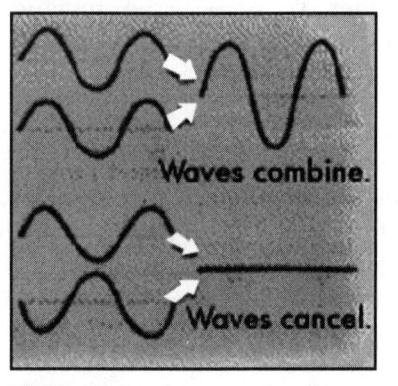

Waves that undergo constructive interference produce a single, stronger wave. Waves that undergo destructive interference tend to cancel each other out.

When you throw a pebble into a pond, you see circular ripples form in the water. But what happens when you throw more than one pebble into the pond at a time? Ripples from the pebbles run into one another and cause interference. When two crests or two troughs meet, the waves combine and form higher crests and deeper troughs. This is called **constructive interference,** and it causes an increase in wave amplitude. If a crest and trough of equal amplitude and frequency meet, they cancel each other out. This is called **destructive interference,** and it causes a decrease in amplitude. See "The Technology of Noise Control" to see how destructive interference can be used to combat unwanted sounds.

Sound Advice

How to Change a Sound

One way to change a sound is to increase or decrease the sound's amplitude. Recall that the amplitude of a sound is related to the amount of energy used to generate the sound waves. If you pluck a guitar string gently, the distance that the string travels in one vibration will be small compared with the distance the string travels in one vibration if you pluck the string forcefully. Changing the amplitude of the vibration changes the loudness of the sound produced by the guitar string. The same thing happens with your vocal chords. When you speak softly, your vocal chords vibrate with a small amplitude. When you speak louder, you increase the amplitude of vibration.

A change in frequency also alters sound. The fewer the vibrations, the lower the frequency and the lower the pitch. Increasing the number of vibrations increases the frequency and raises the pitch of the sound. When you sing low musical notes, your vocal chords are vibrating fewer times per second than they do when you sing high notes.

Turn That Down!

Noise pollution is sound that harms human health. High-energy sounds usually have high volume. **Volume,** or loudness, is measured in **decibels (dB).** Sound energy approximately doubles for every 3 dB increase in loudness. A normal conversation has a volume of about 60 dB, and a rock concert typically has a volume of about 120 dB. The 60 dB increase in loudness from the conversation to the rock concert corresponds to an increase in sound energy by a factor of 1,000,000. Sounds of 90 dB or greater cause permanent hearing damage if exposure to the sound is continuous, and sounds of 170 dB can cause total deafness.

Workers exposed to excessive noise—in factories and on airport runways, for example—must take precautions to protect their hearing. Often you see workers wearing large headphones that protect their ears from loud noises in the workplace. But you don't have to work on a runway to be at risk for hearing damage. Many sounds we encounter in our day-to-day lives can pose a threat to our hearing. Lowering the volume of the TV or your headphones, using ear plugs at rock concerts, and avoiding unnecessary exposure to loud machinery and traffic will help protect you from potential hearing loss.

The Technology of Noise Control

Because of the physical dangers caused by noise pollution, it is desirable to combat offending sounds. Have you ever looked at the ceiling in a busy restaurant? Often the ceiling tiles are specially designed to absorb as much sound as possible. **Passive noise control** is a method of absorbing sound waves or insulating an area from sound. Ear plugs are a common example of passive sound control. Most forms of passive noise control reduce medium and high frequency sounds but are less effective at reducing low frequency sounds.

Another way to combat noise pollution is by active noise control. **Active noise control** uses destructive interference to reduce the amplitude of a sound wave—you actually make more sound to control the offending sound. A speaker is set up to produce sound waves that are the mirror image of the unwanted noise. When the sound waves from the speaker meet the sound waves from the source of noise, the waves cancel each other out. The process works best at low frequencies and in close quarters, such as a small room or an air duct or in headphones.

Exploration 7
Teacher's Notes

In the Spotlight

Key Concepts	A filter is a colored lens that absorbs certain wavelengths of light and transmits others. Different colors of light can be produced by mixing the primary colors of light.
Summary	Ms. Iris Kones is the director of a small community theater. She is preparing for the theater's first big production, which will require many different-colored spotlights. The theater is equipped with two spotlights and 12 colored filters. Ms. Kones wants to know how many different colors of light she can create using the available filters.
Mission	Enlighten the director of a community theater about how to create different colors of light.
Solution	Thirteen colors of light can be produced from the available filters. These colors are as follows: white, red, blue, yellow, green, magenta, cyan, burnt orange, lemon lime, medium blue, raspberry, vivid violet, and pale green.
Background	The lighting design is a crucial component of a stage production. Stage lights can establish the atmosphere or mood of a scene, illuminate key portions of the stage, emphasize dramatic action, and even produce special effects. Lights can also enhance the set of a production by projecting a scenic background onto a screen. Most productions require a range of lighting requirements. For example, one scene may take place outdoors at sunset and another scene may take place indoors at midday. Appropriate lighting is necessary to capture the essence of both situations.
	Different kinds of lights serve different purposes. For example, spotlights are positioned above and behind the audience and direct a concentrated beam of light onto the stage. Background lights are positioned closer to the stage and provide overall color and light to the entire stage. Lights used for background lighting are often a series of lights attached to a beam above the stage. Each light on the beam can be aimed at a different area of the stage to ensure the best coverage. Other pieces of lighting equipment also enhance the lighting design. For example, colored filters can be used to create different colors of light, and various lenses can intensify or soften the brightness of a light. Most stage lights are controlled by a lighting board that has dimmer switches; these switches can be preset at specific levels for the appropriate scenes. Most lighting boards are computerized so that the board adjusts automatically with the push of a button.

Exploration 7 Teacher's Notes, continued

Teaching Strategies

Encourage students to work efficiently through this Exploration. Students can reduce time spent in the lab by detecting color combinations quickly and by noting that testing each filter on one spotlight with every filter on the other spotlight is not necessary. For example, red and cyan light will produce white light no matter which spotlight is red and which is cyan. You may wish to explain to students what complementary colors are. The complementary color of any primary color of light is the color formed by combining the two other primary colors. The complement of blue is yellow (red light plus green light); the complement of red is cyan (blue light plus green light); and the complement of green light is magenta (red light plus blue light). Two complementary colors produce white light when mixed. For example, green light plus magenta light produces white light.

As an extension of this Exploration, you may want to conduct some activities that demonstrate color-vision effects, such as successive contrast. Cut out various shapes from colored construction paper. Have students stare at the colored images for about 30 seconds, and then have them look at a white surface. They will see an afterimage that has the same shape as the original image but with a different color. For example, if the original image was red, the afterimage will be green; and if the original image was blue, the afterimage will be yellow.

Bibliography for Teachers

Peterson, Ivars. "Butterfly Blue." *Science News,* 378(19): November 4, 1995, pp. 296–297.

Winters, Jeffrey. "Vanishing Electrons." *Discover,* 17(7): July 1996, p. 38.

Bibliography for Students

"Electronic Seeing Eye." *Popular Mechanics,* 176(6): June 1996, p. 16.

Bulla, Clyde Robert. *What Makes a Shadow?* Harper Collins, 1995.

Other Media

Learning All About Light & Lasers
CD-ROM for MAC or MS-DOS
Queue, Inc.
338 Commerce Drive
Fairfield, CT 06432
203-335-0906
800-232-2224

In addition to the above CD-ROM, students may find relevant information about light and color by exploring the Internet. Interested students can search for articles using keywords such as *colors of light, visible spectrum,* and *lasers.* Students may also find information about *stage lighting* and *lighting design* on the Internet.

Name _____ Date _____ Class _____

Exploration 7 Worksheet

In the Spotlight

1. Ms. Kones is a little in the dark about the lighting design for her first stage production. What questions do you need to answer for her?

   _____
   _____
   _____
   _____
   _____
   _____
   _____

2. Describe the equipment that Dr. Labcoat has set up on the front table.

   _____
   _____
   _____
   _____
   _____

3. How do colored filters work to create different colors of light? (If you aren't sure, check out the CD-ROM articles.)

   _____
   _____
   _____
   _____
   _____
   _____
   _____
   _____
   _____

Name _____ Date _____ Class _____

Exploration 7 Worksheet, continued

4. As you try different combinations of filters, record your results, including the color equations for each color, in the table below.

Filter 1	Filter 2	Resulting color	Color equation

Name _____ Date _____ Class _____

Exploration 7 Worksheet, continued

5. What are the primary colors of light?

   _____
   _____

6. What are the secondary colors of light?

   _____
   _____

7. What happens when you mix two primary colors of light?

   _____
   _____
   _____

8. What happens when you mix two secondary colors of light?

   _____
   _____
   _____

9. What differences do you notice between mixing the colors of paint on the back counter and mixing colors of light? (If you aren't sure, check out the CD-ROM articles.)

   _____
   _____
   _____
   _____
   _____
   _____
   _____
   _____

Record your answers in the fax to Ms. Kones.

Name _____ Date _____ Class _____

Exploration 7
Fax Form

FAX

To: Ms. Iris Kones (FAX 409-555-2017)

From:

Date:

Subject: In the Spotlight

How many different colors of light (including white light) can be produced using Ms. Kones's two spotlights and twelve filters? _____

Write the color equation for each color that you produced.

_____
_____
_____
_____
_____
_____
_____
_____
_____
_____
_____
_____
_____
_____
_____
_____

Name _____ Date _____ Class _____

Exploration 7 Fax Form, continued

Describe how colored filters produce different colors of light.

Please explain the difference between mixing colors of paint and mixing colors of light.

EXPLORATION 7 • IN THE SPOTLIGHT   213

> **In the Spotlight**
>
> The following articles can also be found by clicking the computer in the CD-ROM laboratory for Exploration 7:
> - *The Nature of Light*
> - *Color Me Beautiful*
> - *Lighten Up!*

**Exploration 7
CD-ROM Articles**

The Nature of Light

The Particle Theory

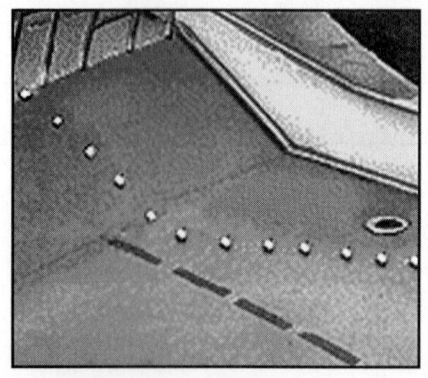

When the ball reaches the bottom of the hill, it changes direction.

Isaac Newton, who is famous for his work with motion, was one of the first scientists to propose the theory that light is composed of particles. He noticed that different colors of light could be produced by shining white light through a prism. The different colors are a result of **refraction**, or the bending of the light as it goes through the prism. Newton argued that this showed that light consists of a stream of particles because the paths of particles also bend when passing through different materials. Consider a ball rolling down a hill. When the ball reaches the bottom of the hill, its path bends, or refracts, and the ball changes direction. Newton suggested that because light also changes its direction, it must be composed of particles.

Newton also noticed that light travels in a straight line, another fact that indicates that light is a stream of particles. An object placed in the path of light casts a shadow; in other words, the object stops the light. In the same way, a barrier placed in the path of a rolling ball stops the ball's motion.

The Wave Theory

Thomas Young demonstrated that light can behave like a wave.

About the same time that Newton proposed his particle theory, a wave theory of light was proposed. The wave theory was not initially accepted because light could not be shown to diffract. **Diffraction** is the bending of waves around barriers. We know that sound waves diffract because we can hear sounds around a corner or outside a doorway. However, we cannot see around a corner, so it does not seem possible that light could be a wave.

However, in 1801, a scientist named Thomas Young conducted an experiment that proved that light does diffract and therefore behaves like a wave. Young made two narrow, parallel slits in a card. He placed a light source on one side of the card and a screen on the other. Young reasoned that if light consists of waves, the light would diffract as it passed through the slits. Young's hypothesis proved true. In this experiment, each slit acts as a separate source of light waves, and the waves coming through the slits **interfere** with each other; that is, crests coincide with crests and troughs coincide with troughs (constructive interference), resulting in bands of light on the screen. When crests coincide with troughs (destructive interference), they cancel each other out and dark bands appear on the screen.

214 HOLT SCIENCE AND TECHNOLOGY INTERACTIVE EXPLORATIONS TEACHER'S GUIDE

The Particle-Wave Theory

Today scientists realize that both the wave model and the particle model are needed to describe all the behaviors of light. This has led to the particle-wave theory of light. How can light exhibit the qualities of both particles and waves? Perhaps a third explanation—the true nature of light—awaits discovery.

Electromagnetic Radiation

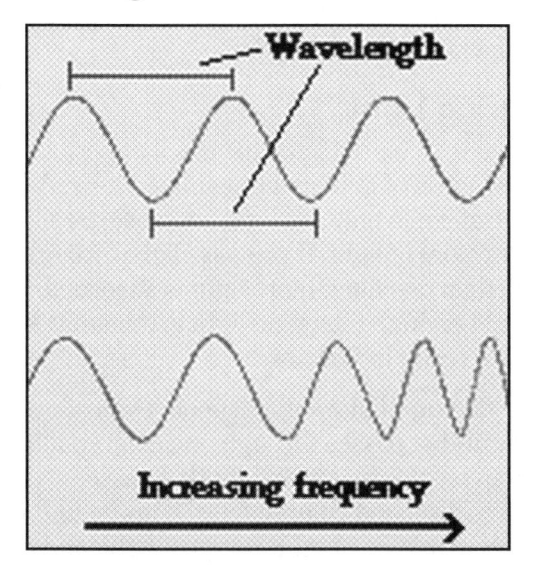

In 1864, James Clerk Maxwell showed that light consists of electromagnetic waves. Electromagnetic waves are waves that carry both electrical and magnetic energy and move through a vacuum at the speed of light (300,000,000 m/s). Heinrich Hertz continued Maxwell's work and showed that electromagnetic waves have a wide range of wavelengths ranging from thousands of kilometers to billionths of a centimeter. Together, these waves make up a type of energy called **electromagnetic radiation,** and they are represented by what we call the electromagnetic spectrum.

When you think of light, you probably think of sunlight, light from lamps, or different colors of light. But these kinds of light are only a part of a much larger phenomenon called the electromagnetic spectrum. The **electromagnetic spectrum** consists of waves of electromagnetic radiation that vary in wavelength and frequency. **Wavelength** is the distance between two corresponding points in a wave, such as from one crest to the next. **Frequency** refers to the number of wavelengths that pass a point during one second.

The Electromagnetic Spectrum

Light that we can see, or **visible light,** is only one part of the electromagnetic spectrum. Visible light contains electromagnetic waves of wavelengths that range from about 400 to 760 nanometers (billionths of a meter). Blue light, for example, has a shorter wavelength and a higher frequency than red light does. When all of the frequencies of visible light are put together, the light appears white.

The types of electromagnetic radiation that we cannot see in the electromagnetic spectrum include radio waves, X rays, infrared light, and ultraviolet light. These types of electromagnetic radiation consist of wavelengths that are either too small or too large for our eyes to see.

Color Me Beautiful

Prisms and Colors of Light

Visible light is only a small portion of the entire range of the electromagnetic spectrum. The color of visible light varies according to its frequency. All the colors of visible light combine to form white light. With a prism, we can see that the process of refraction separates white light into its component colors. Light of shorter wavelengths, such as blue light, bends more than light of longer wavelengths, such as red light. When white light separates, the colors always appear in the same order, from longest wavelength to shortest wavelength. This separation of white light into different colors of light also occurs in nature. You can see it happening every time you see a rainbow. A rainbow results when water vapor functions like a prism and bends the white light from the sun.

Colors of Objects

What makes a red apple red? Why can you say a blue shirt is actually every color *but* blue? Objects appear to be certain colors because white light is shining on the objects and the objects are absorbing, transmitting, or reflecting specific portions of visible light. The color of an object is due to the color of light that the object reflects. For example, if all frequencies of visible light are reflected from an object, the object appears white. If all frequencies of light are absorbed, the object appears black. A red apple looks red because it reflects red light and absorbs all light that is not red. You can say that a blue shirt is every color *but* blue because it absorbs every color of light but blue.

Pigments

If you were to add a drop of red food coloring to a glass of milk, the resulting mixture would be pink. The food coloring contains red pigments that add color to the milk. **Pigments** are basically colored substances that have been processed so they can be used to color other substances. A blue crayon is blue because blue pigment has been added to the crayon wax. The blue pigment reflects only blue light. Colored paint is also a type of pigment.

Primary Colors

We call certain colors primary because all other colors can be made from these colors. When we talk about primary colors, we usually specify whether we are referring to the **primary colors of light** or the **primary colors of paint.** The primary colors of light are red, green, and blue. When you mix these colors together, you get white light. By mixing the primary colors of light in different combinations, you can create all other colors of light, including cyan, magenta, and yellow.

The primary colors of paint are red, yellow, and blue. All other colors of paint can be made from these primary colors. Artists can have an incredibly colorful palette just by combining these three colors. Think about what happens when you mix blue, red, and green paint together. Do you get the same result as when you mix these colors of light (white)? No, because the primary colors of light are different from the primary colors of paint. You can find out why by reading the next section.

Color Formation

When you mix various colors of light, you are adding different wavelengths of light together and "building" white light. This is called **additive color formation.** White light is the presence of all colors of light. Blackness is the absence of all colors of light.

Mixing colors of paint is different from mixing colors of light. Suppose you have a white piece of paper. That piece of paper is reflecting all colors of light. If you paint a splotch of red paint on the paper, the red paint will absorb all the colors of white light except for red light, which it reflects. If you mix paint of another color with the red paint, the resulting blend absorbs more colors of light from white light than either paint would absorb separately. Thus, mixing paint is called **subtractive color formation** because it tends to subtract colors of light from white light. Red, blue, and yellow are the primary colors of paint because when they are mixed together, black, the absence of light, results.

Filtering Light

A **filter** is a colored lens that absorbs certain wavelengths of light and transmits others. A red filter transmits only red light and absorbs all other colors of light. If you pass light through more than one filter, more light is absorbed. If you add enough filters, no light is transmitted and the result is darkness.

What do you think would happen if you allowed the light that passes through one colored filter to overlap on a screen with light that passes through a second colored filter? This would allow you to mix, or add, colors of light, and new color combinations would arise. Using filters that are the primary colors of light (red, blue, and green) allows you to create all other colors of light.

Lighten Up!

Holograms

Holograms are images that are made using laser light. In a holographic camera, a laser beam is split in two. One beam strikes the subject of the image and is reflected onto a piece of film. The second beam hits the film without striking the subject. The two beams interact to produce a microscopic interference pattern that is recorded on the film. In the proper lighting, the image produced by the film appears three-dimensional.

Lighting Systems

Creating the lighting for concerts, plays, and movie sets requires a solid understanding of both additive and subtractive color formation. Light designers use a variety of colored filters to create their spotlight designs. Light designers also consider the color of the backdrop and stage. If the backdrop is colored and the designer uses colored light, subtractive color formation will affect what the audience sees. For example, a blue light will make red, yellow, green, and orange objects look black because they reflect only red, yellow, green, and orange light, respectively, and absorb all other colors of light, including blue.

DNA Pawprints

Key Concepts	DNA is the chemical in chromosomes that carries the genetic information that instructs the cells of living organisms. DNA can be manipulated to produce a DNA fingerprint, which can be used for identification.
Summary	Ms. Jean Poole breeds and shows border collies. She wants to enter three of her younger dogs in an upcoming show, but she lacks the necessary information to complete the dogs' pedigrees. She cannot remember which of her older male dogs sired each of her younger dogs. Ms. Poole has sent blood samples from the young dogs, their mother, and their possible fathers to Dr. Labcoat. Ms. Poole needs to know which older male fathered which young dog, and she would also like some information about the test used to determine their fathers.
Mission	Help a dog breeder obtain the necessary information for her dogs' pedigrees.
Solution	King sired Sugar and Merlin, and Roy sired Domino. The process used to determine the heredity of the young dogs is called DNA fingerprinting.
Background	Puppies of a litter born of purebred dogs share many traits, from size and coloring to temperament. At dog shows, breeders pay careful attention to the parents and offspring of prize-winning animals. Top dog breeders will breed prize-winning parents in the hopes of producing prize-winning offspring. Successive breeding does not always produce desirable results. For example, about 30 percent of all purebred Rottweilers inherit a recessive trait that causes a condition known as *hip dysplasia,* a deformity of the ball-and-socket joint where the hip bone meets the leg bone. Even a mild case of hip dysplasia disqualifies a dog from competition. In the worst cases, it leaves a dog crippled and in pain. Only an artificial hip can eliminate the problem.
	Competitive breeders are aware of the high rate of hip dysplasia among purebreds. Dogs that have the deformity are prevented from breeding. Why, then, are there still so many puppies born with the disorder? First, many dogs that carry the recessive allele that causes the disorder do not show outward signs of hip dysplasia. When two of these carrier dogs breed, some of their puppies can receive two copies of the recessive allele. These puppies inherit the trait for hip dysplasia. Another factor that contributes to the widespread inheritance of hip dysplasia is *inbreeding.* Related prize-winning dogs are frequently bred together, and their puppies are often sold to other breeders. The genetic material from just a few parent dogs recombines again and again, and the chance of two dogs carrying the same recessive gene increases as offspring from closely related litters breed. Therefore, the probability that offspring will inherit both recessive alleles increases.

Exploration 8 Teacher's Notes, continued

Teaching Strategies

As students work through this Exploration, remind them that although they are comparing DNA samples by comparing bands of DNA fingerprints, DNA is not actually a strand of bands. Refer students to Transparency #3 on the back counter in the lab for an actual depiction of the shape of DNA. You may also wish to reinforce among students the idea that a living organism that reproduces sexually receives half its genetic information from one parent and the other half from the other parent. This may help students understand why the DNA fingerprints of each young dog display half Bella's bands and half the appropriate father's bands.

As an extension of this Exploration, you may encourage students to compare the DNA fingerprints of Domino, Merlin, and Sugar. One way to do this would be to sketch the bands of each dog's DNA fingerprint on the chalkboard so that students can compare the bands. Questions to ask students include: How closely related to each other are the young dogs? *(Domino and Merlin are more closely related than are Domino and Sugar; Merlin and Sugar are more closely related than are Sugar and Domino.)* Are there any bands that all three young dogs have in common? *(All three young dogs have Bella's last band in common.)* Merlin and Sugar have the same mother and father; do they also have the same DNA fingerprints? *(No; Merlin has one of Bella's bands and one of King's bands that Sugar does not have, and Sugar has one of Bella's bands and one of King's bands that Merlin does not have.)*

Bibliography for Teachers

Derr, Mark. "The Making of a Marathon Mutt." *Natural History,* 105 (3): March 1996, p. 34.

Jaroff, Leon. "Keys to the Kingdom." *Time,* 148 (14): Fall 1996, p. 24.

Bibliography for Students

McCaig, "The Dogs that Go to Work, and Play, All Day—for Science." *Smithsonian,* 27 (8): November 1996, p. 126.

Mueller, Larry. "Reproducing Junior." *Outdoor Life,* 165 (2): August 1995, p. 62.

Other Media

Biotechnology
Video
National Geographic Society
Educational Services
P. O. Box 98019
Washington, D.C. 20090-8019
800-368-2728

In addition to the above video, students may find relevant information about genetics and electrophoresis by exploring the Internet. Interested students can search for articles using keywords such as *DNA, DNA fingerprinting, chromosomes, genes,* and *genetic engineering.* Students may also find interesting information about the *Human Genome Project* on the Internet.

Name _____ Date _____ Class _____

Exploration 8 Worksheet

DNA Pawprints

1. Ms. Jean Poole wants to enter her dogs in an upcoming dog show. What does she need to know in order to complete the pedigrees for her dogs?

2. What does DNA have to do with inherited characteristics? (If you're not sure, check out the CD-ROM articles.)

3. Explain how DNA fingerprinting works. (Hint: Check out the wall chart.)

Name _____ Date _____ Class _____

Exploration 8 Worksheet, continued

4. Describe the setup on the front table in the lab.

_____

_____

_____

_____

5. Conduct DNA fingerprinting for each young dog, mother, and possible father, and record your results in the table below.

Mother	Young dog	Possible father	Observations of DNA fingerprints
Bella			
Bella			
Bella			
Bella			
Bella			
Bella			
Bella			
Bella			
Bella			

Name _____ Date _____ Class _____

Exploration 8 Worksheet, continued

6. How does Bella's DNA fingerprint compare with the DNA fingerprints of Domino, Merlin, and Sugar?

7. How can you tell which older male fathered each young dog?

8. Look at the material on the lab's back counter. What does it tell you about where DNA is located?

Record your answers in the fax to Ms. Poole.

Name _____ Date _____ Class _____

Exploration 8
Fax Form

FAX

To: Ms. Jean Poole (FAX 512-555-8163)

From:

Date:

Subject: DNA Pawprints

Please indicate which male sired Domino, Merlin, and Sugar.

Domino	○ Duke	○ King	○ Roy
Merlin	○ Duke	○ King	○ Roy
Sugar	○ Duke	○ King	○ Roy

Describe the test that you used to determine which sire matched each of the young dogs.

_____
_____
_____
_____
_____
_____
_____
_____
_____
_____
_____
_____
_____
_____
_____

EXPLORATION 8 • DNA PAWPRINTS

Name _____ Date _____ Class _____

Exploration 8 Fax Form, continued

For Internal Use Only

Please answer the following questions for my laboratory records. Scientists must always keep good records Dr. Crystal Labcoat

> How much genetic information did each young dog inherit from each of its parents? How do you know?
>
> _____
> _____
> _____
> _____
> _____
> _____
> _____
> _____
> _____
> _____
> _____

DNA Pawprints

The following articles can also be found by clicking the computer in the CD-ROM laboratory for Exploration 8:

- *Blueprints for Life*
- *DNA Fingerprints*
- *Advances in Genetic Research*

Exploration 8
CD-ROM Articles

Blueprints for Life

Genes and DNA

When talking about where you get your looks, you may say, "It's in my genes." **Genes** are responsible for the way characteristics are passed on from generation to generation. We inherited our genes from our parents, who inherited their genes from their parents. But what are genes, and how do they determine characteristics such as hair color, eye color, and height? A certain chemical called **DNA** *(deoxyribonucleic acid)* is responsible for the makeup of our genes. DNA is found in almost every one of your cells, and it acts as a set of instructions for how you look and how you inherit or pass on your physical traits. DNA is the blueprint, and you are the finished product.

A molecule of DNA has the structure of a double helix, which resembles a twisted ladder. Each side of the ladder is a separate strand of DNA. Strands of DNA are made up of long chains of smaller units called nucleotides. A **nucleotide** has three parts—a sugar molecule, a phosphate group, and a nitrogen base. The nucleotides differ from one another by the type of nitrogen base present. The four nitrogen bases differ in shape, and their names are adenine (A), thymine (T), guanine (G), and cytosine (C).

Nucleotides are bonded together in a specific way to form the double helix. The sides are formed by bonding the sugar of one nucleotide to the phosphate of the next nucleotide in a continuous chain. Each nitrogen base on one side is bonded to a nitrogen base from the other side to form the rungs of the ladder. The bonded pairs are called **base pairs.** Each base can be paired with only one other base: adenine (A) always pairs with thymine (T), and guanine (G) always pairs with cytosine (C). The order or sequence of the base pairs makes up a gene.

The order of the bases determines what kinds of proteins are made, when they are made, and where they are made. Proteins are responsible for all of the differences in traits among living things. They determine characteristics such as eye color, hair texture, and even how tall you will grow. Although all DNA is made up of the same basic components (a sugar, a phosphate group, and a nitrogen base), not all living things share the same genes. Only in identical twins can you expect to find identical arrangements of nucleotides along a DNA molecule. The millions of arrangements of nucleotides in strands of DNA account for the vast differences in our genetic makeup.

Genes and Chromosomes

Genes occupy specific places on **chromosomes,** which are rod-shaped structures in the nucleus of a cell of a living organism. Chromosomes contain thousands of genes. Humans have 23 pairs of chromosomes in most body cells. As a result of meiosis, sex cells (sperm cells and egg cells) contain only 23 chromosomes apiece—one chromosome from each pair. During sexual reproduction, a sperm cell and an egg cell unite, and the offspring receives 23 chromosomes from the sperm cell (male parent) and 23 chromosomes from the egg cell (female parent) for a total of 46 chromosomes (or 23 pairs). Thus the offspring receives half of its genetic makeup from the mother and the other half from the father.

EXPLORATION 8 • DNA PAWPRINTS 225

Passing Traits Along

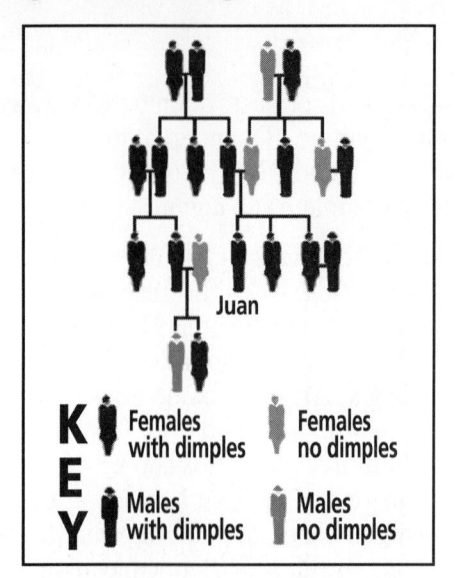

KEY:
- Females with dimples
- Females no dimples
- Males with dimples
- Males no dimples

(Pedigree chart showing Juan)

What do a dimpled chin, a widow's peak, and green eyes have in common? They are all genetic **traits** that can be passed from one generation to the next. Not all genetic traits show up in consecutive generations, though. Why not? The answer lies in the way genes are expressed.

Remember that genes are located on chromosomes. In a given pair of chromosomes, genes that occupy corresponding locations on each chromosome are called gene pairs, and they code for the same trait. Each member of a gene pair is called an **allele**. Alleles are alternative forms of a gene. In a single gene pair, the alleles may or may not be different. For example, a gene pair that codes for the shape of your hairline might have two different alleles—one that codes for a rounded hairline and one that codes for a widow's peak. Conversely, the gene pair could consist of two alleles for a rounded hairline, or it could consist of two alleles for a widow's peak. Often a trait such as eye or hair color is determined by the interaction of many different gene pairs and alleles.

A person possesses two alleles for each trait. Because humans reproduce sexually, one allele comes from the female parent, and the other comes from the male parent. For example, a mother may carry an allele for a widow's peak while a father may carry an allele for a rounded hairline. The offspring will inherit both alleles of this gene for hairline shape. But what determines which form of the gene will be expressed?

The presence of an allele that codes for a specific trait does not ensure that the trait will be expressed in the individual who carries it. Certain alleles are **dominant**; that is, when the allele is present, it masks the effect of a **recessive** allele. If a person inherits one or more dominant alleles, the dominant trait will be expressed. A person must inherit two recessive alleles for the recessive trait to be expressed. When offspring receive one dominant and one recessive allele, they are carriers of the recessive allele. Although the recessive trait is not generally expressed, the allele can be passed along to future generations. This explains why some traits skip a generation.

DNA Fingerprints

How Can DNA Leave a Fingerprint?

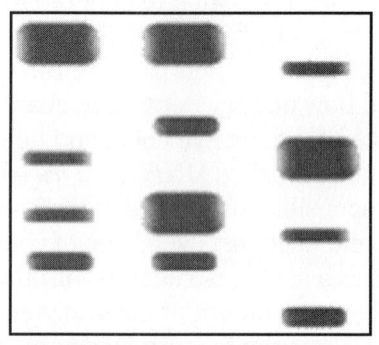

DNA fingerprints can distinguish individuals within a group.

Like fingerprints taken from a person's hand, a DNA fingerprint helps to identify an individual from members of a large population. While taking a thumbprint is a relatively easy procedure, taking a DNA fingerprint requires intensive lab work. First, a sample of DNA must be obtained, usually from blood, hair, or skin. Then the DNA is chemically cut into fragments, and the fragments are sorted by size. Some of these fragments are then tagged with a radioactive probe. When an X ray is taken of the DNA fragments, the fragments appear as a pattern of bands. These patterns allow specialists to determine if two or more DNA samples came from the same person, people who are related, or people who are not related at all.

Cutting DNA Into Fragments

The first step in analyzing DNA involves separating DNA from the rest of the cell parts. The fluid or tissue to be analyzed is transferred to a vial and placed in a centrifuge (a machine that spins material very quickly and causes the DNA to settle at the bottom of the vial). By then heating the DNA, the bonds between the nitrogen base pairs are broken and the strands of DNA come apart. Once the strands have separated, they are chemically cut into fragments. The DNA is then ready for electrophoresis, the process used to separate the DNA fragments.

Electrophoresis

Electrophoresis is a process that uses an electric current to separate the DNA fragments. Two or more samples of DNA are placed in a specially prepared gel tray, and an electric current is passed through the gel tray from one end to the other. The electric current sorts the DNA by size. Because DNA has a negative electrical charge, it will move toward the positive electrode of the electrical source. DNA fragments of different sizes separate because small fragments move toward the positive electrode more quickly than large fragments do. The gel containing the DNA can then be prepared for further examination.

Southern Blot

The Southern Blot is the process that allows us to see the separated fragments of DNA created by the electrophoresis process. The Southern Blot produces a series of bands that allows you to compare one sample of DNA with another. The gel containing the separated fragments of DNA are chemically bathed so that they can be transferred to a nylon membrane. The membrane is then baked to permanently attach the DNA to it. After another series of chemical baths, fragments of DNA are marked with a radioactive tag, and an X ray of the membrane is taken. The DNA will show up on X-ray film as bands of different widths.

Who Needs to Fingerprint DNA?

DNA fingerprinting has a number of applications. For instance, police departments and forensic specialists can use DNA fingerprinting to determine whether a criminal suspect was present at the scene of a crime. Cells containing the DNA to be tested can come from hair, blood, skin, saliva, or other bodily substances. Missing persons may also be tracked using their DNA profile. If an individual's parentage is unclear, DNA fingerprints may be taken to help determine close family relationships. This is done by examining bands of possibly related parents and offspring. Bands in a child's DNA fingerprint that do not match bands in the mother's DNA fingerprint must match bands in the father's DNA fingerprint. In one particularly interesting case, a community in England is planning to create a database of the DNA profiles of all of its 30 dogs. Apparently, the community is having problems with dogs defecating on the sidewalks. The DNA taken from hair samples will be used to determine which dogs are responsible for the mess.

Advances in Genetic Research

The Human Genome Project

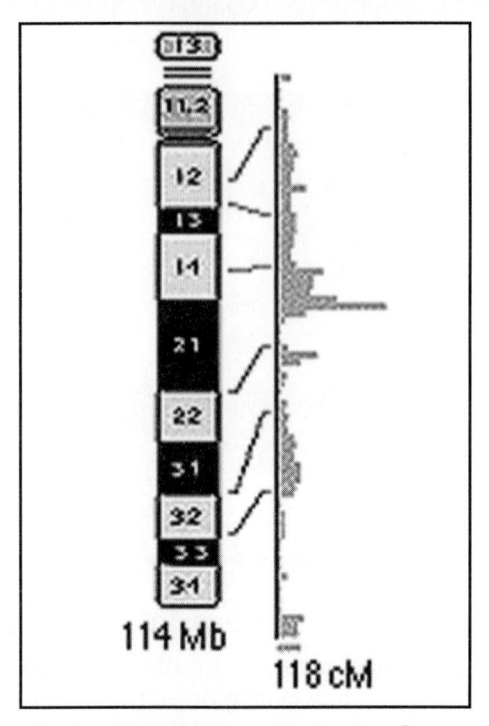

This genetic map shows the density of genes along a given chromosome.

Scientists estimate that humans have over 100,000 different genes. The Human Genome Project is a worldwide effort to analyze all of the genetic material in humans and identify where every gene is located along our 23 pairs of chromosomes. In addition, scientists working on the Human Genome Project plan to learn the base sequence (the order of the bases) for the entire human genome. The **human genome** contains all the genes that make up the master blueprint of a person. The U.S. Human Genome Project began in October 1990. By 1995, over 75 percent of the human genome had been mapped. Researchers plan to use this information to learn more about how genetic mutations play a role in many of today's most common diseases, such as heart disease, diabetes, and birth defects. One day, scientists may be able to treat genetic diseases by correcting errors in the gene itself.

A Gene for Parkinson's Disease

As a result of further genetic research, scientists have been able to pinpoint the location of genes associated with many health disorders. In November 1996, scientists discovered a gene associated with Parkinson's disease. Parkinson's is a progressive disorder that usually strikes adults later in life. Symptoms of this disease include shaky hands, muscular stiffness, and slowness of movement. Mutations (changes in the base-pair order) in the gene will cause classical Parkinson's symptoms. Scientists plan to study this gene in Parkinson's research to help make early diagnoses and to develop new methods of treatment for people currently diagnosed with Parkinson's.

Answer Keys

Holt Science and Technology Interactive Explorations Teacher's Guide

Disc 1 Contents

Exploration 1	231
Exploration 2	232
Exploration 3	235
Exploration 4	237
Exploration 5	238
Exploration 6	240
Exploration 7	242
Exploration 8	245

Disc 2 Contents

Exploration 1	247
Exploration 2	249
Exploration 3	251
Exploration 4	253
Exploration 5	254
Exploration 6	256
Exploration 7	258
Exploration 8	261

Disc 3 Contents

Exploration 1	263
Exploration 2	264
Exploration 3	266
Exploration 4	268
Exploration 5	269
Exploration 6	272
Exploration 7	274
Exploration 8	277

Exploration 1 Worksheet

Something's Fishy

1. What kinds of problems is Mr. McMullet having with his African Cichlids?
   Answers should include details about the condition of the fish, such as their color, their low-energy behavior, and the fact that some are dying.
   (Recommended 5 pts.)

2. What are five variables that might be affecting the African Cichlids?
   a. **Frequency of feeding**
   b. **Presence of ornamental driftwood**
   c. **Variation in temperature**
   d. **Age of the filter**
   e. **Amount of light**
   (Recommended 10 pts.)

3. What will you use for a control as you conduct your investigations?
   One of the two tanks, left undisturbed with settings that remain constant
   (Recommended 5 pts.)

4. Why is this control necessary?
   To establish a comparison with the experimental tank, in which variables can be isolated and the results of changing those variables can be observed
   (Recommended 10 pts.)

5. Would it be better to test one variable at a time or several variables at once? Why?
   It is better to test one variable at a time, because this is the only way of isolating the variable responsible for the change in the cichlids.
   (Recommended 10 pts.)

6. Form a hypothesis for each of the experiments you conduct.
   Hypothesis 1: **Hypotheses should include a reasonable estimation of how each variable might or might not affect the health of cichlids.**
   (Recommended 30 pts.)

EXPLORATION 1 • SOMETHING'S FISHY 3

Hypothesis 2: _____

Hypothesis 3: _____

Hypothesis 4: _____

Hypothesis 5: _____

7. Record your observations as you investigate each hypothesis.

Hypothesis	Observations
1	**Observations should include any changes in the health, behavior, and color of the cichlids.** *(Recommended 20 pts.)*
2	
3	
4	
5	

8. Were your experiments faulty in any way? If so, what steps did you take to correct your experiments?
   Errors may include failure to conduct research before beginning an experiment, trying to change the control tank, and testing more than one variable at a time.
   (Recommended 10 pts.)

Record your conclusions in the fax to Mr. McMullet.

4 HOLT SCIENCE AND TECHNOLOGY INTERACTIVE EXPLORATIONS TEACHER'S GUIDE

Exploration 2 Worksheet

Shut Your Trap!

1. What are Ms. Lily N. Lotus and the Bogs Are Beautiful Appreciation Society concerned about? **Answers should include a summary of the society's concerns, including a description of why the flytraps are disappearing from the wild. Examples may include poaching and a shrinking natural habitat.** *(Recommended 5 pts.)*

2. What are three variables that may be affecting the growth of the Venus' flytraps?
   a. **Light**
   b. **Humidity**
   c. **Plant food** *(Recommended 10 pts.)*

3. What will you use for a control in your investigations? **One of the two terrariums, left undisturbed with settings that remain constant** *(Recommended 5 pts.)*

4. There are 24 possible variable settings for the experimental terrarium. However, you have only enough flies to conduct 10 experiments. What steps can you take to make sure that you find a solution before you run out of flies? **The order in which variables are tested can quickly reduce the number of experiments necessary. For instance, testing the effect of the plant food first will eliminate half of the possible number of experiments.** *(Recommended 20 pts.)*

5. Form a hypothesis for how each variable affects the growth of the Venus' flytraps.
   Hypothesis 1: **Hypotheses should include a reasonable estimation of how variables may or may not combine to affect the growth of the flytraps.** *(Recommended 20 pts.)*
   Hypothesis 2: _____
   Hypothesis 3: _____

EXPLORATION 2 • SHUT YOUR TRAP 11

Exploration 1 Fax Form

FAX

To: Mr. Ray McMullet (FAX 512-555-8633)
From: _____
Date: _____
Subject: African Cichlids

What is your recommendation? **Remove the ornamental driftwood from your tanks; it is harming the cichlids.** *(Recommended 10 pts.)*

✂ -

For Internal Use Only

Please answer the following questions for my laboratory records. Scientists must always keep good records Dr. Crystalobscot

During your experiments, which ONE of the following changes had a positive effect on the fish?

EXPERIMENTAL VARIABLES

☐ FEED FISH   ☐ INCREASE TEMPERATURE
☐ TURN LIGHT OFF   ☐ CHANGE FILTER
☒ REMOVE ORNAMENTAL DRIFTWOOD

(Recommended scoring: 50 points for above answer; 30 points for that response plus another; and 0 points for not indicating the above response.)

Please explain why the African Cichlids responded to the above change. **African Cichlids require water that is alkaline. The driftwood that was added to the tanks was slowly leaching tannins into the water. The tannins caused the water to be too acidic for the cichlids. Removing the driftwood corrected the pH level in the tank.** *(Recommended 25 pts.)*

What effect did the change that you made have on the fish? **The fish became active and colorful.** *(Recommended 15 pts.)*

EXPLORATION 1 • SOMETHING'S FISHY 5

---

232 HOLT SCIENCE AND TECHNOLOGY INTERACTIVE EXPLORATIONS TEACHER'S GUIDE

6. Record your observations in the table below as you investigate each hypothesis.

Plant food	Humidity	Hours of light	Observations
Yes	0%	5	**Observations should include specific changes in the growth, health, or blooming of the flytraps.** *(Recommended 30 pts.)*
	0%	10	
	0%	15	
	0%	20	
	25%	5	
	25%	10	
	25%	15	
	25%	20	
	50%	5	
	50%	10	
	50%	15	
	50%	20	
No	0%	5	
	0%	10	
	0%	15	
	0%	20	
	25%	5	
	25%	10	
	25%	15	
	25%	20	
	50%	5	
	50%	10	
	50%	15	
	50%	20	

7. Were your experiments faulty in any way? If so, what steps did you take to correct them?

Errors may include failure to conduct research before beginning the experiments, incorrectly setting the variables on the control terrarium, or testing variables at random.

(Recommended 10 pts.)

Record your conclusions in the fax to Lily N. Lotus.

What effect did the plant food have on the plants? Why?

Plant food cannot be given to the Venus' flytrap because the plant receives nitrogen from insects, the extra nutrients provided by plant food will kill the plant. *(Recommended 25 pts.)*

Exploration 2
Fax Form

FAX

To: Lily N. Lotus (FAX 910-555-5657)

From:

Date:

Subject: Optimal Growing Conditions for Venus' Flytraps

What is your recommendation? **Optimal growing conditions for the Venus' flytrap include 15 hours of light, 50% humidity, and no plant food.** *(Recommended 5 pts.)*

✂ — ✂ — ✂ — ✂ — ✂ — ✂ — ✂ — ✂ — ✂ — ✂ — ✂ — ✂

For Internal Use Only

Please answer the following questions for my laboratory records. Scientists must always keep good records *Dr. Crystalclear*

During your experiments, which values proved to be optimal for the Venus' flytrap?

Optimal Values (select one per row)   **EXPERIMENTAL VARIABLES**

5	10	**15**	20	HOURS OF LIGHT PER DAY
0	25	**50**		PERCENT HUMIDITY
YES	**NO**			ADD PLANT FOOD?

(Recommended scoring: 50 points if optimal values are indicated as shown; 30 points for 10 hours of light, 50% humidity, and no plant food; and 30 points for 20 hours of light, 50% humidity, and no plant food. All other answers should receive 0 points.)

What effect did the hours of light per day have on the plants? Why?

The Venus' flytrap produces a healthy seedpod in 90 days when exposed to 15 hours of light per day. *(Recommended 10 pts.)*

What effect did the percent humidity have on the plants? Why?

The Venus' flytrap produces a healthy seedpod in 90 days when exposed to 50% humidity. *(Recommended 10 pts.)*

5. Use the table below to record your observations of each slide of microorganisms. Make sure that you write out the name of each microorganism in the left-hand column.

	Protista	Observations
1.	*Euglena*	Transparent green; single celled; one flagellum
2.	*Paramecium*	Transparent blue; single celled; with cilia
3.	*Amoeba*	Reddish brown; single celled; uses cytoplasm as pseudopodia
4.	Algae, dinoflagellate	Gray and pink; two flagella
	Monera	Observations
5.	Round-shaped bacteria, *Staphylococcus aureus*	Red and green; single celled; round shape
6.	Rod-shaped bacteria, *Escherichia coli*	Orange with green spots; single celled; oblong shape
7.	Spiral-shaped bacteria, *Spirillum*	Transparent purple; single celled; spiral shape
	Fungi	Observations
8.	Mildew, *Peronospora manschuriea*	Black, curving fibers
9.	Mold, *Rhizopus stolonifer*	Black spots on yellowish, hairlike fibers
10.	Yeast, *Saccaromyces cerevisiae*	Blue, round, interconnected bubbles

(Recommended scoring: 2 points for each correct name and 2 points for each appropriate description.)

Exploration 3 Worksheet

Scope It Out!

1. What does Dr. Viola Russ need to know about the ancient microorganisms?
   Dr. Russ needs to know the classification of the ancient microorganisms and information about the likely role they played in the life of the ancient bee. *(Recommended 5 pts.)*

2. What does Dr. Russ intend to do with the results?
   She will use this information to produce new and more effective antibiotics.
   (Recommended 5 pts.)

3. What will you use to conduct your investigation?
   Making a visual comparison of the ancient microorganisms and the modern microorganisms is the only way to make an identification, so the slides should be carefully analyzed under the microscope. Answers should also emphasize researching the CD-ROM articles for clues about the ancient microorganisms' function. *(Recommended 10 pts.)*

4. What do the ancient microorganisms look like under the microscope?
   The ancient microorganisms are reddish pink, single celled, and have an oblong shape.
   (Recommended 10 pts.)

Exploration 3
Fax Form

FAX

To: Dr. Viola Russ (FAX 805-555-2266)

From:

Date:

Subject: Ancient microorganism classification and probable function

What are the ancient microorganism's classification and function? *The ancient microorganism is a type of rod-shaped bacteria that belongs to the kingdom Monera. The organism probably aided the bee in digestion and in fighting disease.*

(Recommended 25 pts.)

For Internal Use Only

Please answer the following questions for my laboratory records. Scientists must always keep good records. *Dr. Crystalobscot*

Which of the following may be used to classify the ancient microorganisms? Place an X in the left-hand column beside the correct answer(s).

KINGDOM	PROTISTA		MONERA		FUNGI
	Euglena	X	Round-shaped bacteria		Mildew
	Paramecium	X	Rod-shaped bacteria		Mold
	Amoeba		Spiral-shaped bacteria		Yeast
	Algae				

(Recommended scoring: 50 points if both kingdom Monera and rod-shaped bacteria are selected; 30 points if only kingdom Monera or only rod-shaped bacteria is selected; and 0 points for all other answers.)

What role did this microorganism most likely play in the life of the ancient bee? *Like Bacillus sphaericus, the ancient microorganism probably aided the bee in digestion and in fighting disease.*

(Recommended 25 pts.)

---

6. Which modern microorganisms look the most like the ancient microorganisms? *The rod-shaped bacteria of the kingdom Monera share the same basic shape as the ancient microorganisms.*

(Recommended 10 pts.)

7. How might you classify the ancient microorganisms and find out more about their likely role in the life of the ancient bee? *Information can be obtained from the CD-ROM articles, as well as from outside references such as books, periodicals, and other materials about microorganisms. Consulting an expert from a local college or university could also be helpful.*

(Recommended 20 pts.)

Record your conclusions in the fax to Dr. Russ.

5. How might you find this information?

The CD-ROM articles are an excellent source of such information. In addition, outside resources such as books, scientific articles, or an expert at a local college, university, or scientific institution may be consulted. *(Recommended 10 pts.)*

6. Record helpful data here, continuing on the back of the page if necessary. **Answers should reflect a careful review of the CD-ROM articles.** *(Recommended 10 pts.)*

7. When you have finished, evaluate the procedure that you used to complete this Exploration. What would change about your procedure? Did you perform any activities that were not useful to you? If so, which ones?

Answers will vary but should include a clear reflection of the steps taken to complete this Exploration. Answers may include the fact that determining the mass and volume of each metal was not useful in solving Dr. Stokes's problem. *(Recommended 10 pts.)*

8. How could you improve your procedure?

Answers will vary but could include the fact that carefully planning an experiment before proceeding is an effective way to avoid problems. *(Recommended 10 pts.)*

Record your conclusions in the fax to Dr. Stokes.

Exploration 4 Worksheet

What's the Matter?

1. What problem does Dr. Stokes need you to help him solve?

The tip of an instrument for measuring lava temperature has melted, and Dr. Stokes needs some help determining what metal to use as a replacement. *(Recommended 10 pts.)*

2. Dr. Labcoat has gathered a set of metal samples for you to analyze. List the name of each metal and record its melting point and boiling point in the following data chart:

Metal	Name of metal	Melting point (°C)	Boiling point (°C)
Cu	Copper	1083	2567
Sn	Tin	232	2270
Pt	Platinum	1722	3827
Ti	Titanium	1660	3287
W	Tungsten	3410	5660
Al	Aluminum	660	2467

(Recommended 30 pts.)

3. How might the information in the table be useful to you in solving Dr. Stokes's problem?

Dr. Stokes's instrument will have to be repaired with a material that is very heat resistant since it will be inserted into hot lava. It is important, therefore, to determine the temperature limits of likely materials. *(Recommended 10 pts.)*

4. What additional information could help you solve the problem?

Answers should include information such as the following: the temperature of molten lava, the availability and expense of each metal, and how difficult it is to obtain each metal. *(Recommended 10 pts.)*

Exploration 5 Worksheet

Element of Surprise

1. Mr. Stamp needs your help. Describe your assignment.
   The assignment is to determine the chemical reactivity of 12 elements to water so that Mr. Stamp can construct the best possible transport containers for the elements and deliver them safely to Antarctica. *(Recommended 10 pts.)*

2. What materials are available in Dr. Labcoat's lab to help you complete your assignment?
   There are samples of 9 of the 12 elements, a beaker of water, a pair of tweezers, a pipette, an eyedropper, and a chemical storage rack. *(Recommended 10 pts.)*

3. Describe what you will do to test each element's reactivity to water.
   A sample of each element should be added to the beaker of water, and then observations can be made about any reaction that takes place. *(Recommended 10 pts.)*

4. Record your findings about each sample in the spaces that follow. *(Recommended 40 pts.)*

   a. barium: **When tested, the water bubbles quickly and turns cloudy.**
   Answers to a–i should include a description of each sample's reaction to water. Students may also have included additional information about each element from the CD-ROM articles.

   b. calcium: **When tested, the water bubbles quickly and turns cloudy.**

Exploration 4 Fax Form

FAX

To: John Stokes, Ph.D. (FAX 080-555-9822)

From:

Date:

Subject: Metal recommendation

What is your recommendation? **Use titanium to replace the tip of the lava analyzer.** *(Recommended 20 pts.)*

✂ · · ✂ · · ✂ · · ✂ · · ✂ · · ✂ · · ✂ · · ✂ · · ✂ · · ✂ · · ✂ · · ✂ · · ✂

For Internal Use Only

Please answer the following questions for my laboratory records. Scientists must always keep good records. — *Dr. Crystalclearcoat*

Please indicate your metal selection here: **Titanium**
(Recommended scoring: 50 points for above answer; 30 points for tungsten or platinum; and 0 points for copper, aluminum, or tin.)

How do the particles of this metal behave during the following phases:

solid? **The particles are packed closely together in a rigid form until about 1659°C.** *(Recommended 10 pts.)*

liquid? **At about 1660°C, the particles begin to separate, vibrating against each other and flowing in a random pattern.** *(Recommended 10 pts.)*

gas? **At about 3287°C, the particles begin to bounce off each other and into the air in a random way.** *(Recommended 10 pts.)*

c. cesium: **When tested, the water turns purple and there is a slightly explosive reaction above the surface of the water.**

d. neon: **When tested, the water bubbles briefly but remains clear.**

e. potassium: **When tested, the sample bubbles on the surface of the water before creating a smoking reaction above the surface. The water changes color.**

f. radon: **When tested, the water bubbles briefly but remains clear.**

g. rubidium: **When tested, the top portion of the water turns purple, and there is a slightly explosive reaction on the surface.**

h. strontium: **When tested, the sample sinks to the bottom of the beaker and bubbles rise quickly toward the surface, changing the color of the water.**

i. xenon: **When tested, the water bubbles briefly but remains clear.**

EXPLORATION 5 • ELEMENT OF SURPRISE   41

---

5. What additional information do you need to complete your assignment (to determine the reactivity of krypton, magnesium, and sodium to water)?

The periodic table of elements in Dr. Labcoat's lab and the information in the CD-ROM articles are needed to determine the reactivity of the remaining elements to water.

(Recommended 15 pts.)

6. Now that you know the reactivity of each of the 12 elements, how do you think Mr. Stamp should pack the chemicals when preparing to deliver them to Antarctica?

Sample answer: It is very important that neither the highly reactive nor the reactive elements come in contact with water. Therefore, the containers that hold the cesium, potassium, rubidium, sodium, barium, calcium, strontium, and magnesium samples should be watertight. The samples of neon, radon, xenon, and krypton will not be damaged if they come in contact with water, so it is less important for those containers to be watertight.

(Recommended 15 pts.)

Record your conclusions in the fax to Mr. Stamp.

42   HOLT SCIENCE AND TECHNOLOGY INTERACTIVE EXPLORATIONS TEACHER'S GUIDE

Exploration 6 Worksheet

The Generation Gap

1. Wendy Powers is a home builder who is considering a plan to make her homes more efficient. What has she asked you to do to help her?

   Ms. Powers has asked for assistance in testing the Electroprop wind turbine to determine if it would be cost-effective to install in the San Francisco area. *(Recommended 4 pts.)*

2. Dr. Labcoat has set up a system that enables you to test the energy output of the wind turbine at eight different speed settings. Run the tests, and record your results below.

Meters per second	Kilowatt-hours	Time-lapse indicator
1	1	7 days
2	8	7 days
3	27	7 days
4	65	7 days
5	125	7 days
6	200	7 days
7	350	7 days
8	500	7 days

(Recommended 16 pts.)

3. What is the value of the above information?

   It shows the amount of energy generated by several speeds of wind in a 7-day period of time.

   (Recommended 10 pts.)

Exploration 5 Fax Form

FAX

To: Mr. Fred Stamp (FAX 011-619-555-7669)

From:

Date:

Subject: Chemical Properties of Elements

Select the appropriate classification for each of the following chemicals:

CHEMICAL REACTIVITY WITH WATER

CHEMICAL	EXTREMELY REACTIVE	REACTIVE	NOT REACTIVE
BARIUM		X	
CALCIUM		X	
CESIUM	X		
NEON			X
POTASSIUM	X		
RADON			X
RUBIDIUM	X		
STRONTIUM		X	
XENON			X

(Recommended scoring: Each correct answer is worth 2 points, for a total of 18 points possible.)

Please utilize the above information to predict the chemical reactivity of the following chemicals:

CHEMICAL	EXTREMELY REACTIVE	REACTIVE	NOT REACTIVE
KRYPTON			X
MAGNESIUM		X	
SODIUM	X		

(Recommended 32 pts.)

(Recommended scoring: 50 points for three correct answers; 30 points for 1 or 2 correct answers; 0 points for no correct answers.)

How did the periodic table help you to answer Mr. Stamp's questions?

Elements in the same column of the periodic table belong to the same chemical group and have similar chemical properties. Knowing how one element in a group reacts with water can be used to predict the reactivity of other elements in that same group.

7. Use the table below to record the energy output of the Electroprop wind turbine and your calculations of the savings it will bring the homeowner.

Wind speed in meters per second (m/s)	Energy output over 7 days in kilowatt-hours (kWh)	Savings per year ($)	Years until Electroprop wind turbine has paid for itself
1	1	7.28	2000
2	8	58.24	250
3	27	196.56	75
4	65	473.20	31
5	125	910.00	16
6	200	1456.00	10
7	350	2548.00	6
8	500	3640.00	4

Record your conclusions in the fax to Ms. Powers.

(Recommended 36 pts.)

---

4. What other information will you need to complete your task?

The average wind speed in the San Francisco Bay area, the cost per kilowatt-hour of electricity in the San Francisco Bay area, how to calculate the amount of money saved by using a wind turbine, and how to calculate the amount of time before the wind turbine pays for itself

(Recommended 10 pts.)

5. Use the lab resources to find this information. You can record your notes here.

Answers should reflect a careful review of the CD-ROM articles.

(Recommended 10 pts.)

6. How will you calculate the amount of money a wind turbine can save a homeowner over the course of a year?

By multiplying the number of kilowatt-hours generated by the wind turbine by the cost per kilowatt-hour to find the amount of savings per week, and then by multiplying the weekly savings by 52 weeks to find the amount of savings per year

(Recommended 14 pts.)

Exploration 7 Worksheet

Teach It While It's Hot!

1. What has Dr. Labcoat asked you to do to help Mr. McCool?

   To make sure that the demonstration she has prepared will be an effective teaching tool for Mr. McCool's lesson about the difference between temperature and heat.

   (Recommended 5 pts.)

2. What information would be helpful to know before you begin your investigation?

   A familiarity with both temperature and heat as well as the units in which each is measured would be useful.

   (Recommended 5 pts.)

3. Where do you think you could find this information?

   In the CD-ROM articles or in a science textbook

   (Recommended 5 pts.)

4. What happens when heat energy is applied to a beaker of water?

   The temperature of the water increases.

   (Recommended 5 pts.)

5. Record your observations as each beaker (quantity) of water is placed on the ring stand.

   a. green (100 mL)

   The water begins to boil after 6 seconds. It takes about 34,000 J of heat energy to bring 100 mL of water from 20°C to 100°C.

   (Recommended 10 pts.)

   b. red (200 mL)

   The water begins to boil after 11 seconds. It takes about 67,000 J of heat energy to bring 200 mL of water from 20°C to 100°C.

   (Recommended 10 pts.)

Exploration 6 Fax Form

FAX

To: Ms. Wendy Powers (FAX 415-555-2766)

From:

Date:

Subject: Wind-Energy Economics

Is it cost-effective to use the Electroprop to generate energy in the San Francisco area? Why or why not?

Yes, because the Electroprop wind turbine will pay for itself in 10 years. After this initial pay-back period, the turbine will save the homeowner approximately $1456 each year for 10 years, for a total savings of $14,560.   *(Recommended 20 pts.)*

For Internal Use Only

Please answer the following questions for my laboratory records. Scientists must always keep good records. Dr. Crystalofcoat

Approximately how much money would the Electroprop save a San Francisco homeowner in an average year?

$4	$28	$140	$400	**$1460**	$2800
1	5	**10**	16	28	250

Approximately how many years would it take for the Electroprop to pay for itself?

(Recommended scoring: 50 points for two correct answers; 30 points for one correct answer; and 0 points for no correct answers.)

How many years would it take for the Electroprop to pay for itself if the average wind speed in the San Francisco area were each of the following:

8 m/s? **4 years**

5 m/s? **16 years**

2 m/s? **250 years**

(Recommended scoring: 10 points for each correct answer.)

c. blue (300 mL)

The water begins to boil after 17 seconds. It takes about 101,000 J of heat energy to bring 300 mL of water from 20°C to 100°C. *(Recommended 10 pts.)*

6. How can you calculate the amount of heat required to increase the temperature of 600 mL of water from 20°C to 100°C?

Sample answer: There is a direct relationship between the mass of a substance and the amount of heat required to raise the temperature of that mass by a specific amount. Since a 600 mL beaker contains six times as much water as a 100 mL beaker, six times as many joules will be required to raise the temperature of 600 mL of water from 20°C to 100°C as are required to raise the temperature of 100 mL of water from 20°C to 100°C.

(Recommended 15 pts.)

7. Why did Dr. Labcoat provide you with three different quantities of water?

To show the direct relationship between the amount (mass) of water and the amount of heat energy needed to raise the water to a specific temperature. For example, it requires three times as much energy to raise 300 mL of water from 20°C to 100°C as it does to raise 100 mL of water from 20°C to 100°C.

(Recommended 10 pts.)

8. Use the graph as well as your knowledge of temperature and heat to describe what this demonstration shows.

Sample answer: This demonstration shows that when heat energy is added to a substance, the temperature of that substance increases. That temperature is an average measure of how "hot" the substance is at that particular instant, whereas, the heat of the substance tells how much total energy the substance contains. The amount of heat required to raise the temperature of a substance to a specific temperature is directly related to the mass of the substance (in this case, the volume of the water). *(Recommended 15 pts.)*

9. Based on what you've learned during this activity, would you recommend this demonstration to Mr. McCool? Why or why not?

Answers will vary, but recommendations should be clear, concise, and well supported with examples. Sample response: Yes. This demonstration effectively shows that there is a direct relationship between the mass of a substance and the amount of heat required to raise the temperature of that mass by a specific amount. The visual demonstration and the graph present the information in a format that is understandable to middle-school students.

(Recommended 10 pts.)

Record your conclusions in the fax to Mr. McCool.

Exploration 7
Fax Form

FAX

To: Mr. Kelvin McCool (FAX 512-555-4328)

From:

Date:

Subject: Teaching Recommendations

What relationship is represented by your graph?

The graph shows the relationship between change in temperature and amount of heat added. The three different lines on the graph represent the three different quantities of water and show the direct relationship between mass and the amount of heat required to cause a specified change in temperature. For example, if the mass of a sample of water is doubled, then the amount of heat required to raise the temperature of that sample of water from 20°C to 100°C will also double.

(Recommended 10 pts.)

Please use your data to determine the answers to the following questions:

Which beaker contains the most heat energy at 100°C?	**GREEN**	**RED**	**BLUE**
Approximately how much heat would have to be added to increase the temperature of 600 mL of water from 20°C to 100°C?	100,000 joules	200,000 joules	300,000 joules

(Recommended scoring: 50 points for two correct answers; 30 points for one correct answer; 0 points for no correct answers.)

What is the approximate temperature of each sample of water when the amount of heat energy added is 30,000 joules?

GREEN: 92°C **RED:** 56°C **BLUE:** 44°C

(Recommended scoring: each correct answer is worth 5 points, for a total of 15 points.)

Please write and answer one essay question that will help Mr. McCool's students understand the relationship between temperature and heat.

Sample question: What is the difference between temperature and heat? Sample answer: Heat is the total amount of energy in a substance, and temperature is a measurement of the average amount of energy in a substance, or how "hot" a substance is. A 100 mL beaker of water and a 500 mL beaker of water that are the same temperature have different amounts of heat energy. The more massive amount of water will contain more heat energy.

(Recommended 25 pts.)

Exploration 8 Worksheet

Flood Bank

1. Ms. Sandy Banks is the chairperson of her local environmental-impact committee. What has she asked you to do to help her?

   Ms. Banks needs to know whether building a dam to create a reservoir will have long-term geologic and environmental effects on the river downstream.

   (Recommended 15 pts.)

2. How can the simulation that Dr. Labcoat has provided help you with your investigation?

   The simulation provides a means of visually comparing the geological effects of natural water flow with the geological effects of controlled water flow. In addition, natural versus controlled flooding can also be tested. *(Recommended 20 pts.)*

3. Conduct your research, recording your notes in the space provided below.

   Notes should include descriptions of a variety of the effects that dams have on rivers, including effects on the riverbanks and the ecosystem as a whole, as well as effects on the flow of sediment. Additionally, notes could include references to the effects of the Glen Canyon Dam on the Colorado River or other information from the CD-ROM articles.

   (Recommended 20 pts.)

4. Use the table below to record the effects of both types of water flow at varying flow rates.

Type of flow	Low	Medium	High
Natural flow	The stream's banks widen slightly, and small amounts of sediment are deposited on the inner curves of the stream.	The stream widens, and the banks become slightly more curved. A large amount of sediment is apparent.	The stream widens, and its banks curve much more sharply. Levels of sediment deposition continue to increase.
Regulated flow	The stream's banks widen slightly, just as they did during natural flow, but little or no sediment is deposited along the banks.	The stream widens, and the banks become slightly more curved, just as they did during natural flow, but little or no sediment is apparent.	The stream widens much more broadly during regulated flow than during natural flow. The curves are less snakelike because there is little or no sediment deposition.

   (Recommended 30 pts.)

5. Compare the results of your stream-table observations with the information you discovered in your research. Which environmental and geological effects of dams are not reproducible by a stream table?

   Several effects of the dam are not represented by the stream table. Examples include the long-term effects of the loss of sediment that the river naturally carries downstream, the temperature difference in the water caused by controlled flooding, and the severe effects on the ecosystem as a whole.

   (Recommended 15 pts.)

   Record your conclusions in the fax to Ms. Banks.

Which volume of water flow has the greatest impact on the formation of these geologic features?

| Low | Medium | **High** |

Is it possible to maintain a healthy river environment downstream from a dam?

| YES | **NO** |

(Recommended scoring for the above two questions: 50 points for two correct answers; 30 points for 1 correct answer; 0 points for no correct answers.)

Why or why not?

Dams are very disruptive to the natural river environment. Even if water could flow through the dam at the same rate it would normally flow, much of the river's sediment would remain on the bottom of the reservoir. Controlled flooding could help keep the river clear of debris and could help distribute some of the sediment below the dam, but sediment above the dam would still need to be moved somehow from behind the dam to areas downstream. Also, some method would need to be devised to increase the temperature of the water before it affected the ecosystems downstream.

(Recommended 25 pts.)

Exploration 8
Fax Form

FAX

To: Ms. Sandy Banks (FAX 520-555-7239)

From:

Date:

Subject: Possible effects of a dam on the natural river environment

What effect does a dam have on the natural river environment?

Because a dam blocks the natural flow of a river, it limits the erosion and deposition of sediment by the river. So instead of collecting and carrying nutrient-rich sediment downstream, the river drops the sediment at the bottom of the reservoir. As a result, the ecosystems downstream are deprived of the sediment that is often the basis of their food chains. Dams also prevent flood waters from periodically scouring the river bottom. As a result, debris settles on the river's bottom and creates barriers in front of inlets where fish would normally spawn. Where flood waters once carved out the river's banks and created new sandbanks farther downstream, vegetation takes root. These changes in the landscape threaten the organisms that live in and along the river system.

(Recommended 25 pts.)

Exploration 1 Worksheet

What's Bugging You?

1. Dr. Mike Roe's patients are suffering from the same symptoms. Describe what Dr. Roe wants you to do so that he can treat his patients.

    Dr. Roe has sent samples of six organisms collected from each of his sick patients. He wants to know the identity of each organism and what kind of relationship each organism has with its host so that he can double-check his diagnosis of his patients' illness.

    (Recommended 10 pts.)

2. What is the difference between the two different trays of slides on the front table in Dr. Labcoat's lab?

    One tray contains slides of the six unidentified sample organisms from Dr. Roe's patients, and the other contains slides of 13 known reference organisms that live in or on the human body. *(Recommended 5 pts.)*

3. How will you use the microscope and the slides to help Dr. Roe?

    The six sample organisms can be identified by comparing each one with the 13 reference slides. *(Recommended 5 pts.)*

4. What other information will you then need to find out about the six sample organisms?

    The function of each organism within the human body and whether the organisms have commensalistic, mutualistic, or parasitic interactions with humans

    (Recommended 10 pts.)

5. How could you find such information?

    By thoroughly researching the CD-ROM articles *(Recommended 4 pts.)*

---

6. Examine the 13 reference slides under the microscope, and record your observations, along with the name of each organism, in the table below.

Name of organism	Observations
1. *Clonorchis sinensis*	oblong with internal, hairlike fibers
2. *Enterobius vermicularis*	wormlike with tapered ends
3. *Schistosoma mansoni*	thin, curly body with two-pronged ends
4. *Naegleria fowleri*	protist-shaped with two flagella
5. *Giardia lamblia*	oval-shaped with two eyespots and multiple flagella
6. *Rickettsia rickettsii*	round bacteria with darker, spotty insides
7. *Propionibacterium acnes*	rod-shaped bacteria
8. *Dermatophagoides farinae*	protist-shaped with clawlike extensions
9. *Staphylococcus epidermis*	round bacteria, clustered
10. *Demodex folliculorum*	oblong-bodied mite
11. *Salmonella*	rod-shaped bacteria with multiple flagella
12. *Entamoeba gingivalis*	amoeba-shaped, with dark "eyespots"
13. *Streptococcus mutans*	two strands of interconnected round bacteria

(Recommended 26 pts.)

Exploration 1
Fax Form

FAX

To: Dr. Mike Roe (FAX 206-555-7272)

From:

Date:

Subject: Identification of Sample Organisms

What are the six organisms taken from Dr. Roe's patients, and what is their classification: mutualistic, commensalistic, or parasitic?

Sample	Name of organism	Mutualistic	Commensalistic	Parasitic
1	*Entamoeba gingivalis*		•	
2	*Enterobius vermicularis*			•
3	*Demodex folliculorum*		•	
4	*Rickettsia rickettsii*			•
5	*Propionibacterium acnes*	•		
6	*Streptococcus mutans*			•

(Recommended scoring: 50 points for 6 correct responses; 30 points for 3 to 5 correct responses; and 0 points for less than 3 correct responses.)

Of the six sample organisms, which is the most likely cause of the patients' fever, chills, and muscle aches? *(Recommended 25 pts.)*

Sample 1	Sample 2	Sample 3	**Sample 4**	Sample 5	Sample 6

What is the name of the illness caused by this organism, and how is it transmitted to humans?

Rickettsia rickettsii, a bacterium that lives in ticks, causes Rocky Mountain spotted fever. This illness can be transmitted when a human is bitten by an infected tick. The bacteria are transmitted through the tick's saliva. *(Recommended 25 pts.)*

82 HOLT SCIENCE AND TECHNOLOGY INTERACTIVE EXPLORATIONS TEACHER'S GUIDE

---

7. Now examine the six sample slides, and record your observations in the table below. Refer to your observations of the reference slides, and use the third column below to record any similarities you find between the reference slides and each sample slide.

Sample	Observations	Similarities to reference slides
1	amoeba-shaped with dark eyespots	looks like reference slide 12, *Entamoeba gingivalis*
2	wormlike with tapered ends	looks like reference slide 2, *Enterobius vermicularis*
3	oblong-bodied mite	looks like reference slide 10, *Demodex folliculorum*
4	round bacteria with darker, spotty insides	looks like reference slide 6, *Rickettsia rickettsii*
5	rod-shaped bacteria	looks like reference slide 7, *Propionibacterium acnes*
6	two strands of interconnected round bacteria	looks like reference slide 13, *Streptococcus mutans*

(Recommended 30 pts.)

8. In the CD-ROM articles, read about the interactions that each of the six sample organisms have with humans. Record your notes here. Use another piece of paper if necessary.

Answers should reflect an ability to determine the type of interaction that exists between humans and each organism based on information about symptoms of infection and the environment of the organisms. *(Recommended 10 pts.)*

Record your answers in the fax to Dr. Roe.

EXPLORATION 1 • WHAT'S BUGGING YOU? 81

248 HOLT SCIENCE AND TECHNOLOGY INTERACTIVE EXPLORATIONS TEACHER'S GUIDE

Exploration 2
Worksheet

Sea Sick

1. Shelley C. Waters has sent some unusual creatures to the lab. What is wrong with them, and what does she want you to do for her?

   The supposedly spectacular creatures are not thriving in their tanks. Ms. Waters wants to know what kind of creatures they are and what their ideal habitat is so that she can feature them as part of a permanent exhibit at the Marine Exploratorium.

   (Recommended 10 pts.)

2. The front lab table is pretty crowded with equipment! Describe the different parts of the setup.

   There are two pressurized tanks (a control tank and an experimental tank) that contain the creatures. Each tank has a pressure reading, a temperature reading, and a description of the mineral composition of the water in each tank. There is also a container of food, a bottle of antibacterial solution, and extra filters for the tanks.

   (Recommended 10 pts.)

3. Why is it necessary to use a control in this experiment?

   A control is necessary to establish a comparison with the experimental tank, in which variables can be isolated. The results of changing those variables can be observed in the appearance of the sea creatures.

   (Recommended 10 pts.)

4. What are the settings on the control tank?

   The control tank has a water temperature of 15°C, a pressure of 1 atm, a sea-water composition of magnesium sulfate, magnesium chloride, and sodium chloride (Solution A). There is no food, no antibacterial water conditioner, and no change in the filter.

   (Recommended 5 pts.)

5. Conduct your experiments, and record your observations in the table below.

Change in variable	Observations
Add food.	*There is no change in the creatures.*
Add antibacterial water conditioner.	*The top portion of the creatures shrank away.*
Increase pressure in tank.	*An increase in tank pressure to 250 atm made the top portion of the creatures thicken, turn pink, and rise up halfway to vertical.*
Change filter.	*There is no change in the creatures.*
Increase temperature.	*There is no change in the creatures.*
Change mineral composition of sea water.	*Changing to Solution B (hydrogen sulfide, sodium chloride, and magnesium chloride) caused the top portion of the creatures to thicken, turn deep red, and rise up to vertical.*

(Recommended 30 pts.)

6. What other information do you need to give Ms. Waters?
   The kind of sea creatures, their ideal habitat, and their nutritional requirements

   (Recommended 5 pts.)

7. Describe how you might learn this information.
   By researching the CD-ROM articles and other resources, such as magazines, a higher level science textbook, or the Internet

   (Recommended 5 pts.)

8. Examine the slides Dr. Labcoat has set up on the back counter. Describe your findings. What you see plays a role in the habitat of the mystery creatures. What is that role? Explain. (Hint: Check out the CD-ROM articles.)
   Sample answer: The slides contain bacteria that are found near hydrothermal vents and that are crucial to the process of chemosynthesis. The bacteria break down the hydrogen sulfide in vent waters into carbon and oxygen atoms that vent organisms, including tubeworms, can use for food.

   (Recommended 10 pts.)

9. Describe the relationship between your experiment results and what you have learned about these organisms and their habitat.
   The natural habitat of the tubeworm, the deep-oceanic zone, is a high-pressure, mineral-rich environment. Thus, the pressure in the exhibit tanks must be increased, and the mineral composition of the sea water in the tanks must be changed to include hydrogen sulfide. The conditions of the tubeworms' natural habitat must be closely replicated if the tubeworms are to be as healthy in captivity as they are in their natural environment.

   (Recommended 15 pts.)

Record your answers in the fax to Ms. Waters.

EXPLORATION 2 • SEA SICK   91

Exploration 2
Fax Form

FAX

To: Ms. Shelley C. Waters (FAX 619-555-8368)

From:

Date:

Subject: Habitat

Name of organisms:
Giant tubeworms

(Recommended 10 pts.)

Where do these animals live?
On the deep-ocean bottom in the vicinity of hydrothermal vents

(Recommended 10 pts.)

During your experiments, which of the following changes had a positive effect on the animals?

	ADD FOOD.
	ADD ANTIBACTERIAL WATER CONDITIONER.
•	CHANGE MINERAL COMPOSITION OF SEA WATER.
•	INCREASE PRESSURE IN TANK.
	CHANGE FILTER.
	INCREASE TEMPERATURE.

(Recommended scoring: 50 points for the above responses only; 30 points for the above responses plus another response; 20 points for one of the above responses plus another response; and 0 points for any other response(s).)

What are the nutritional requirements of these animals?
Since the giant tubeworms get their food energy from the bacteria that live in their tissues, hydrogen sulfide must be present in the water so that the bacteria can carry out chemosynthesis.

(Recommended 30 pts.)

92   HOLT SCIENCE AND TECHNOLOGY INTERACTIVE EXPLORATIONS TEACHER'S GUIDE

Exploration 3 Worksheet

Moose Malady

1. Mr. Oleson is very concerned about some moose in western Sweden. What does he want to know?

   He wants to know what is making the moose sick and why the numbers of blueberries are declining.

   (Recommended 10 pts.)

2. What has Mr. Oleson told you about both the new and the traditional habitat and niche of the moose?

   Traditionally, the moose live in a boggy terrain and eat blueberries. However, since the numbers of blueberries have been declining recently, the moose have wandered into nearby fields, where they are now eating barley and oats. *(Recommended 5 pts.)*

3. Mr. Oleson had the internal organs from several of the moose that died tested. What did he learn?

   He learned that the moose's internal organs had high concentrations of molybdenum and low concentrations of copper, cadmium, and chromium. *(Recommended 10 pts.)*

4. Describe the equipment that Dr. Labcoat has set up on the front table.

   There are three test tubes; one contains rainwater from the region, one contains blueberries from the preserve, and one contains barley and oats from the farmers' fields. Twelve indicator solutions and a pH indicator are also set up on the table.

   (Recommended 10 pts.)

5. What is the purpose of an indicator solution?

   An indicator solution can show the presence and level of concentration of a particular substance in a solution. For example, an aluminum indicator can show how concentrated aluminum is in a sample solution such as rainwater. *(Recommended 10 pts.)*

6. Conduct your analysis of Mr. Oleson's samples, and record your results in the table below.

Test		Results		
Indicator solution	**Sample 1**	**Sample 2**	**Sample 3**	
Aluminum (Al)	Normal	High	Normal	
Cadmium (Cd)	Normal	High	Low	
Chromium (Cr)	Normal	Normal	Low	
Copper (Cu)	Normal	Normal	Low	
Iron (Fe)	Normal	High	Low	
Manganese (Mn)	Normal	High	Low	
Molybdenum (Mo)	Normal	Low	High	
Nitrogen (N)	High	High	High	
Phosphorus (P)	Normal	Normal	Normal	
Potassium (K)	Normal	Normal	High	
Sulfur (S)	High	High	High	
Zinc (Zn)	Normal	High	Low	
pH indicator				
pH level	4.0	4.4	7.8	

(Recommended 25 pts.)

7. In terms of your results above, what do *High*, *Normal*, and *Low* refer to?

   How concentrated a given mineral is, with High indicating an unusually large concentration of a particular mineral *(Recommended 10 pts.)*

8. What clues do you get from analyzing the pH levels of the sample solutions? The CD-ROM articles should help you with your analysis.

   The rainwater is acidic (pH of 4.0), so this area in western Sweden is experiencing acid rain. The acid rain has probably affected the blueberries, as is evident by their pH of 4.4. Since the barley and oats have a significantly higher pH of 7.8, the field in which they grow has probably been treated with lime. *(Recommended 20 pts.)*

Record your answers in the fax to Mr. Oleson.

Which two tests provided the strongest evidence to support your answers to the previous questions?

Test selected	Best evidence
Aluminum	
Cadmium	
Chromium	
Copper	X
Iron	
Manganese	
Molybdenum	
Nitrogen	
Phosphorus	
Potassium	
Sulfur	
Zinc	
pH level	X

(Recommended scoring: 50 points for both correct responses; 30 points for one of the two correct responses; and 0 points for neither correct response.)

What can Mr. Oleson do to help restore the moose's health?

Answers may vary but should be clear and logical. One possible solution: Supply the moose with food grown in fields that have not been treated with lime so that the moose obtain adequate levels of copper.   *(Recommended 20 pts.)*

102   HOLT SCIENCE AND TECHNOLOGY INTERACTIVE EXPLORATIONS TEACHER'S GUIDE

Exploration 3

FAX

To: Mr. Hans Oleson (FAX 46-18-555-5374)

From:

Date:

Subject: Mysterious Moose Malady

What is making the moose sick, and what is causing the numbers of blueberries to decline? Explain.

The moose are sick because the barley and oats they have been eating are low in copper. The crops are low in copper because the farmers in the area have been spreading lime on their fields to counter the effects of acid rainfall. The liming process causes molybdenum to dissolve out of the soil particles where it may be absorbed by plants. High levels of molybdenum in a plant keep it from absorbing much copper, which is also present in the soil. A certain amount of copper is necessary for life processes to be properly carried out. Without adequate amounts of copper, the moose can suffer the symptoms described by Mr. Oleson. The numbers of blueberries are declining because acid rain has increased the level of acidity in the soil. Blueberry bushes do not grow well in highly acidic soil.

(Recommended 30 pts.)

EXPLORATION 3 • MOOSE MALADY   101

252   HOLT SCIENCE AND TECHNOLOGY INTERACTIVE EXPLORATIONS TEACHER'S GUIDE

Exploration 4 Worksheet

Force in the Forest

1. Mr. Solimões has asked you to help him improve the efficiency of his packing facility. What information, specifically, does he require?

   Mr. Solimões needs to know what amount of force is required to get crates of specific masses all the way to the end of the 30 m track without falling off and damaging their contents.

   (Recommended 10 pts.)

2. The name of Mr. Solimões's company is Sustainable Rain Forest Products, Inc. What are sustainable rain-forest products, and why might they be important?

   Sample answer: Sustainable rain-forest products are native rain-forest products that are grown so that there is very little waste or damaging of the raw materials. Sustainable businesses grow no more crops than they can sell, and they do not harm the natural resources when harvesting these crops. These combined efforts aim to ensure that the natural resources can provide for future generations.

   (Recommended 15 pts.)

3. Describe the lab setup on the front table of Dr. Labcoat's lab.

   There is a pneumatic pushing device to push crates of different masses (5 kg, 10 kg, 25 kg, and 50 kg) along a roller track. The track has a sharp curve at about the halfway point.

   (Recommended 10 pts.)

4. How does this lab setup compare with Mr. Solimões's production line? What is it designed to accomplish?

   The lab setup is a scaled-down version of the 30 m roller track that Mr. Solimões wants to use. It is designed to simulate the actual results that Mr. Solimões could expect if crates of 5 kg, 10 kg, 25 kg, or 50 kg were pushed with various forces by the pneumatic pushing device along the track.

   (Recommended 15 pts.)

5. Conduct all of the possible tests with the lab equipment, and record your results in the table below.

Mass (kg)	Pushing force (N)	Observations
5	80	The crate goes all the way to the end of the track, and the light goes on.
5	160	The crate falls off the track.
5	400	The crate falls off the track.
5	800	The crate falls off the track.
10	80	The crate does not get all the way to the end of the track.
10	160	The crate goes all the way to the end of the track, and the light goes on.
10	400	The crate falls off the track.
10	800	The crate falls off the track.
25	80	The crate does not get all the way to the end of the track.
25	160	The crate does not get all the way to the end of the track.
25	400	The crate goes all the way to the end of the track, and the light goes on.
25	800	The crate falls off the track.
50	80	The crate does not get all the way to the end of the track.
50	160	The crate does not get all the way to the end of the track.
50	400	The crate does not get all the way to the end of the track.
50	800	The crate goes all the way to the end of the track, and the light goes on.

(Recommended 30 pts.)

6. What other force (besides the pushing force) influences the behavior of the crate on the track? Describe how this force works. Explore the CD-ROM articles if you aren't sure of the answer.

   Sample answer: Frictional force is caused by the friction between the crate and the track. This force opposes the initial force and works in the opposite direction in which the crate is moving. The more massive the moving object is, the stronger the frictional force is. That's why the most massive box moves the least distance when pushed with a force of 80 N.

   (Recommended 20 pts.)

Record your answers in the fax to Mr. Solimões.

Exploration 5 Worksheet

Extreme Skiing

1. Mr. Ludwig Guttman needs your help. What information has he requested?
   He wants to know the best material to use in the manufacture of a prosthetic limb for a ski racer. The material must be strong and lightweight. Students may also mention that it should be corrosion resistant.
   (Recommended 10 pts.)

2. Examine the devices on the front table and the back counter in Dr. Labcoat's lab. What are they designed to do?
   The device on the front table is designed to test the strength of materials by applying compressive, tensile, and shear forces. The devices on the back counter are designed to find the mass and volume of materials.
   (Recommended 5 pts.)

3. What are the differences between compressive, tensile, and shear forces? If you aren't sure, explore the CD-ROM articles.
   Compressive forces are the forces that cause material to become pressed together; tensile forces are the forces that cause a material to stretch; and shear forces are the forces that tend to cause a material to tear or be cut apart.
   (Recommended 15 pts.)

4. Conduct all the necessary tests with the device on the front table in the lab, and record your results in the table below.

Amount of Force Each Material Can Withstand (10^6 N/m²)

Material	Compressive	Tensile	Shear
Aluminum alloy	400	400	400
Carbon fiber	1050	1050	250
Magnesium alloy	250	250	15
Oak	80	80	5
Steel	500	500	250
Titanium alloy	1000	1000	250
Tungsten alloy	1000	1000	150

(Recommended 25 pts.)

Exploration 4 Fax Form

FAX

To: Mr. Gustavo Solimões (FAX 55-11-707-8988)

From:

Date:

Subject: Forces in the Rain Forest

What is a newton?
A newton is the international metric (SI) unit used to measure force. Abbreviated N, one newton (1 N) is approximately equal to the amount of force required to cause a 1 kg mass to accelerate at a rate of 1 m/s².
(Recommended 25 pts.)

How much force must be applied to the following masses to successfully send each to the end of the 30 m roller track?

MASS (kg)	FORCE (N)				
5	80	160	400	800	
10	80	160	400	800	
25	80	160	400	800	
50	80	160	400	800	

(Recommended scoring: 50 points for 4 correct responses; 25 points for 2 or 3 correct responses; and 0 points for less than 2 correct responses.)

How much force must be applied to a 100 kg mass to send it the full length of a 30 m roller track? How much force must be applied to a 200 kg mass?
100 kg mass: 1600 N; 200 kg mass: 3200 N
(Recommended 25 pts.)

Exploration 5 — Fax Form

FAX

To: Mr. Ludwig Guttman (FAX 404-555-3296)
From:
Date:
Subject: Material Recommendation

Please indicate your material selection here: **Carbon fiber**
(Recommended scoring: 50 points for carbon fiber; 30 points for titanium alloy or aluminum alloy; and 0 points for oak, magnesium alloy, steel, or tungsten alloy.)

Why did you recommend this material?
Carbon fiber can withstand the greatest compressive and tensile forces. It can also withstand reasonably strong shear forces, and it has a very low density. Carbon fiber is also corrosion resistant. A prosthesis made from carbon fiber would be extremely strong, lightweight, and corrosion resistant.
(Recommended 20 pts.)

Please complete the following chart:

MATERIAL	Compressive 10^6 (N/m²)	Tensile 10^6 (N/m²)	Shear 10^6 (N/m²)	Density g/mL
Aluminum alloy	400	400	400	2.70
Carbon fiber	1050	1050	250	2.00
Magnesium alloy	250	250	15	2.00
Oak	80	80	5	0.80
Steel	500	500	250	8.00
Titanium alloy	1000	1000	250	4.50
Tungsten alloy	1000	1000	150	19.00

(Recommended 30 pts.)

---

5. Why do you think the figures for shear strength are usually lower than those for compressive or tensile strength?
Answers will vary, but students should recognize that shear forces oppose the natural grain of a material, whereas compressive and tensile forces press or stretch a material along the grain.
(Recommended 10 pts.)

6. What other information do you need to know before recommending a material to Mr. Guttman?
Students need to find out which of the strongest materials is also the most lightweight. They can also determine which one is most corrosion resistant.
(Recommended 5 pts.)

7. How can you determine this information?
Use the equipment on the back counter in the lab, and research the CD-ROM articles.
(Recommended 5 pts.)

8. Complete the following table:

Material	Mass (g)	Volume (mL)	Density (g/mL)
Aluminum alloy	108	40	2.70
Carbon fiber	80	40	2.00
Magnesium alloy	80	40	2.00
Oak	32	40	0.80
Steel	320	40	8.00
Titanium alloy	180	40	4.50
Tungsten alloy	760	40	19.00

(Recommended 15 pts.)

9. What do you notice about the volume of each material? Why might this be the case?
The volumes are equal. This is so that the densities and, consequently, the masses of the materials can be compared. If the volumes were different, then a larger mass could be due to a larger amount of the material, rather than due to a physical property of the material itself.
(Recommended 10 pts.)

Record your answers in the fax to Mr. Guttman.

Rock On!

1. Claudia Stone is supervising the development of an information kiosk at a new state park. What does she need to know to complete the project?

   Ms. Stone needs to know whether the rock specimens from the park area are igneous, sedimentary, or metamorphic, and she needs to know where each rock type fits in the rock cycle.

   (Recommended 10 pts.)

2. What kinds of things are available in the lab to help you provide Ms. Stone with the information she needs?

   There are 10 rock specimens, a magnifying glass, and a rock-cycle chart. The CD-ROM articles will also be useful in answering Ms. Stone's questions. *(Recommended 5 pts.)*

3. Briefly describe the differences between igneous, sedimentary, and metamorphic rocks. Use the CD-ROM articles to help you.

   Igneous rocks are produced by the cooling and solidification of magma or lava; sedimentary rocks are produced from cemented mineral particles; and metamorphic rocks are rocks that have undergone change as a result of intense heat and pressure.

   (Recommended 10 pts.)

4. As you examine each of the rock specimens, record your notes for each rock in the table below. Be sure to show the steps you took through the Identification Key. For example, conglomerate is (1a) grainy and made of more than one material → (2b) particles held together by natural cement → sedimentary.

Rock Classification Table

Rock	Name	Steps from identification key	Classification
1	Obsidian	(1b) The rock is made of only one material. (5a) The rock is glassy.	Igneous
2	Gabbro	(1a) The rock is grainy. (2a) The particles are interlocking. (3b) The grains are of two or more different types. (4a) The grains are arranged randomly.	Igneous
3	Granite	(1a) The rock is grainy. (2a) The particles are interlocking. (3b) The grains are of two or more different types. (4a) The grains are arranged randomly.	Igneous
4	Limestone	(1a) The rock is made of more than one material. (2b) The particles are held in place by a natural cement.	Sedimentary
5	Marble	(1a) The rock is grainy. (2a) The particles are interlocking. (3a) All of the grains are of the same type.	Metamorphic

Rock Classification Table, continued

Rock	Name	Steps from identification key	Classification
6	Quartzite	(1a) The rock is grainy. (2a) The particles are interlocking. (3a) All of the grains are of the same type.	Metamorphic
7	Sandstone	(1a) The rock is grainy. (2b) The particles are held in place by a natural cement.	Sedimentary
8	Basalt	(1b) The rock is made of only one material. (5a) The rock is porous.	Igneous
9	Shale	(1a) The rock is grainy. (2b) The particles are held in place by a natural cement.	Sedimentary
10	Slate	(1b) The rock is made of only one material. (5b) The rock is made of strong, flat sheets.	Metamorphic

(Recommended scoring: 2 points each for the correct classification steps, and 2 points for the correct rock type, for a total of 40 possible points)

5. What other information do you need to give Ms. Stone?

A description of the relationship that exists between igneous, sedimentary and metamorphic rocks

(Recommended 5 pts.)

EXPLORATION 6 • ROCK ON! 127

---

6. How could you find this information?

By examining the rock cycle chart in the lab and by researching the CD-ROM articles

(Recommended 5 pts.)

7. Which rock specimens were the easiest to classify? the most difficult to classify? Give reasons for your choices.

Answers will vary, but students should provide at least one clear example of each. For example, the glassy appearance of obsidian and the visible sheets within slate make these rocks fairly easy to classify. The difficulty in distinguishing the types of grains in gabbro and marble may make these rocks more difficult to classify. In addition, students may have difficulty recognizing the difference between interlocking grains and grains that are held together with natural cement.

(Recommended 15 pts.)

8. Examine the samples on the back counter of the lab. What is unusual about them, and what is the explanation?

They glow in the ultraviolet light because they are fluorescent rock specimens.

(Recommended 10 pts.)

9. What other information could you recommend that Ms. Stone provide in the information kiosk?

Answers will vary but other information could include where the types of rocks tend to be located, what percentage of the rocks in the park is represented by each kind of rock, and what this information reveals about when and how the area developed. The kiosk could also highlight the presence of the interesting fluorescent rocks.

(Recommended 5 bonus pts.)

Record your answers in the fax to Ms. Stone.

Exploration 7 Worksheet

Space Case

1. You've been asked to help Estelle de la Luna. What is your mission? Describe the function of the equipment in the lab, and give Ms. de la Luna some instructions for how to use it.

(Recommended 10 pts.)

2. Describe the different parts of the equipment on the front table in the lab.

Sample answer: The equipment consists of a scaled-down model of the Sun, Earth, and Moon. The Earth is on an arm that looks like it could revolve around the Sun. The Moon is also on an arm that looks like it could revolve around the Earth.

(Recommended 10 pts.)

3. What happens when you click on the equipment? Describe the setup now.

Sample answer: The view shifts to an overhead view, the model of the Sun lights up, and half of the Earth's surface is lit by the Sun. There are eight numbered positions of the Moon around the Earth, and half of the Moon's surface is lit by the Sun at all times.

(Recommended 10 pts.)

4. What do you think the numbered squares represent?

The numbers represent the eight phases of the Moon.

(Recommended 5 pts.)

EXPLORATION 7 • SPACE CASE 135

Exploration 6 Fax Form

FAX

To: Ms. Claudia Stone (FAX 719-555-0612)

From:

Date:

Subject: Rock Classification

Please classify specimens 1 through 10.

Specimen number	Specimen name	Classification		
		Sedimentary	Igneous	Metamorphic
1	obsidian		•	
2	gabbro		•	
3	granite		•	
4	limestone	•		
5	marble			•
6	quartzite			•
7	sandstone	•		
8	basalt		•	
9	shale	•		
10	slate			•

(Recommended scoring: 50 points for 10 correct responses; 30 points for 6 to 9 correct responses; and 0 points for less than 6 correct responses.)

Describe the relationship that exists among sedimentary, igneous, and metamorphic rocks.

Sample answer: Any of these rock types can be converted to any other type through the rock cycle. For example, igneous, metamorphic, or sedimentary rocks can be eroded and deposited to form sediments that can harden into sedimentary rock. Sedimentary rock can be buried, compressed, and heated to form metamorphic rock or, if heated more intensely, igneous rock. Metamorphic rock, if melted, will harden into igneous rock. Igneous rock, if compressed and heated, will become metamorphic rock.

(Recommended 50 pts.)

EXPLORATION 6 • ROCK ON! 129

5. Why do you think the squares are numbered counterclockwise?

The Moon revolves around the Earth in a counterclockwise direction.

(Recommended 10 pts.)

6. In the following chart, describe what the Moon looks like at each numbered location that you click:

Position number	Description of Moon
1	*The Moon is not visible.*
2	*A sliver of the right side of the Moon's face is visible. It looks like a crescent.*
3	*The right half of the Moon's face is visible.*
4	*Almost the whole face of the Moon is lit up. Only a piece of the Moon's face is in darkness.*
5	*The Moon is full; its entire face is lit up.*
6	*Almost the whole face of the Moon is lit up. Only a piece of the right side of the Moon's face is in darkness.*
7	*The left half of the Moon's face is visible.*
8	*A sliver of the left side of the Moon's face is visible. It looks like a crescent.*

(Recommended 30 pts.)

7. Dr. Labcoat wants you to identify the name of each of the numbered positions on the model. What are they? (Hint: Check out the CD-ROM articles.)

Positions 1 through 8, respectively, are new Moon, waxing crescent Moon, first quarter Moon, waxing gibbous Moon, full Moon, waning gibbous Moon, last quarter Moon, and waning crescent Moon.

(Recommended 10 pts.)

8. What should you tell visitors about the scale of this model?

Sample answer: The model is not built to scale. For example, the Earth is actually much smaller in comparison with the Sun, and it is also farther away from the Sun than is implied by this model.

(Recommended 10 pts.)

9. What other astronomical phenomena could this model demonstrate?

Sample answers include solar and lunar eclipses, the orbits of the Earth and the Moon, and the seasons.

(Recommended 5 pts.)

Record your answers in the fax to Ms. de la Luna.

Please write brief, easy-to-follow directions explaining how to use this equipment.

Answers will vary based on whether students describe the model as an actual physical model of the Sun, Earth, and Moon, or as a computerized abstraction of a physical model, which is actually what they see in the computer laboratory. Accept either method of describing the model. Sample answer: Use this equipment to model the phases of the Moon and the relative positions of the Sun, Earth, and the Moon during the Moon's orbit around Earth. When you select a numbered position on the model, the appropriate phase of the Moon as seen from Earth can be viewed.

(Recommended 30 pts.)

How much of the Moon's surface is sunlit during each of the eight phases?

Half of the Moon's surface is sunlit at all times (except during a lunar eclipse), although not all of the lit surface is always visible from Earth.

(Recommended 10 pts.)

FAX

To: Estelle de la Luna (FAX 520-555-6666)

From:

Date:

Subject: Space Case

What does the planetarium model demonstrate?

The model demonstrates the cause of the Moon's phases.

(Recommended 10 pts.)

Determine the correct name for each of the eight numbered positions on the model.

	Position 1	Position 2	Position 3	Position 4	Position 5	Position 6	Position 7	Position 8
first quarter			•					
full Moon					•			
last quarter							•	
new Moon	•							
waning crescent								•
waning gibbous						•		
waxing crescent		•						
waxing gibbous				•				

(Recommended scoring: 50 points for all 8 correct responses; 30 points for 5 to 7 correct responses; and 0 points for less than 5 correct responses.)

Exploration 8 Worksheet

How's It Growing?

1. Rosie Flores needs your gardening expertise. What does she want to know?

   She wants to know why some hydrangeas, which were grown from the cuttings of a hydrangea that produces blue flowers, are now producing pink flowers.

   (Recommended 10 pts.)

2. How did the reader that Ms. Flores is responding to grow his hydrangeas? Describe this process. (Hint: Check out the CD-ROM articles.)

   Vegetative reproduction is the process of growing a new plant from either the stem or the leaf cuttings of an existing plant. The new roots can then be transplanted to a larger pot or into a garden plot.

   (Recommended 12 pts.)

3. Dr. Labcoat has set out some materials on the front table in the lab. Describe her setup.

   There are 2 hydrangeas; 2 lights; some distilled water, plant food, and powdered bauxite; a container of ladybugs; a time-lapse indicator; a pH indicator and a thermometer in the soil of each plant; and 2 magnifying glasses.

   (Recommended 10 pts.)

4. Why is there a control plant and an experimental plant?

   A control plant allows for a comparison with the experimental plant. During the experiment, variables can be isolated, and the results of changing those variables can then be more easily observed.

   (Recommended 10 pts.)

5. Record the settings you choose for your control setup.

   Answers will vary.

   (Recommended 10 pts.)

EXPLORATION 8 • HOW'S IT GROWING? 145

---

6. Conduct all of the necessary experiments, recording your observations of how each of the variables below affects the hydrangeas.

   a. Hours of light per day

   This variable affects the temperature of the soil. With 8 hours of light per day, the soil is 21°C, with 10 hours, it is 24°C, and with 12 hours, it is 27°C.

   (Recommended 12 pts.)

   b. Soil conditioner

   This variable affects the pH of the soil and, as a result, the color of the flowers. Adding powdered bauxite lowers the soil pH, and the plant produces blue flowers.

   (Recommended 12 pts.)

   c. Plant food

   This variable affects the number of flowers that the plant produces. The plant produces three flowers without plant food and five flowers with plant food.

   (Recommended 12 pts.)

   d. Ladybugs

   This variable affects the presence of aphids on the leaves of the plants. When ladybugs are introduced, the aphids disappear.

   (Recommended 12 pts.)

Record your answers in the fax to Ms. Flores.

Exploration 8
Fax Form

FAX

To: Ms. Rosie Flores (FAX 213-555-0612)

From: _____

Date: _____

Subject: How's it growing?

How can two hydrangea plants that are genetically identical produce different-colored flowers?

The color of a hydrangea's flowers is not genetically determined. Changing the acidity of the soil causes changes in the color of the plant's blossoms. In acidic soil, hydrangea blossoms are blue. In neutral or alkaline soil, hydrangea blossoms are pink.

(Recommended 10 pts.)

For Internal Use Only

Please answer the following questions for my laboratory records. Scientists must always keep good records. *Dr. Crystalofscent*

During your experiments, which one of the following variables helped you to discover the correct answer to the above question?

	CHANGE HOURS OF LIGHT PER DAY.
X	ADD SOIL CONDITIONER.
	ADD PLANT FOOD.
	INTRODUCE LADYBUGS.

(Recommended scoring: 50 points for the above response; 30 points for the above response with any other response; and 0 points for any response but the one above.)

What effects did each of the following variables have on the hydrangeas? Please explain.

a. Hours of light per day

Changing the hours of light per day changed neither the number of blossoms nor the color of the blossoms, but it did affect the temperature of the soil.

(Recommended 10 pts.)

EXPLORATION 8 • HOW'S IT GROWING? 147

---

b. Soil conditioner

Adding soil conditioner caused the pH of the soil to drop and the color of the blossoms to change from pink to blue. *(Recommended 10 pts.)*

c. Plant food

Adding plant food increased the number of blossoms.

(Recommended 10 pts.)

d. Ladybugs

Adding ladybugs changed neither the number of blossoms nor the color of the blossoms, but the ladybugs did get rid of the aphids on the plant.

(Recommended 10 pts.)

148 HOLT SCIENCE AND TECHNOLOGY INTERACTIVE EXPLORATIONS TEACHER'S GUIDE

5. Use the equipment to conduct the necessary tests, and record your data in the table below.

Test tube	Test-tube contents	Time to diffuse (sec.)
A	perfume	15
B	rotten eggs	5
C	garlic	12
D	alcohol	6
E	cinnamon	10

(Recommended 15 pts.)

6. Why are the temperature and pressure kept constant for this experiment? (If you're not sure, check out the CD-ROM articles.)

Sample answer: Temperature and pressure must be kept constant so that the rate of diffusion of each chemical is not affected by different variables. Temperature and pressure both influence the rate of diffusion of a substance. Particles at higher temperatures and pressures diffuse faster because of the increased amount of kinetic energy of the particles.

(Recommended 15 pts.)

7. What does the equipment on the back counter in Dr. Labcoat's lab demonstrate?

Sample answer: This equipment shows how a higher temperature increases the rate of diffusion. The 30°C water allows the ink particles to diffuse faster than the 10°C water does.

(Recommended 10 pts.)

8. How do you smell an odorous chemical? Use the CD-ROM articles to help you explain how your sense of smell works.

Sample answer: After you inhale odor particles that are diffused through the air, some of the odor particles settle on olfactory receptor cells in the nose. These receptor cells send a message along the olfactory nerve to the brain, where the smell is interpreted.

(Recommended 10 pts.)

Record your answers in the fax to Ms. Foushen.

Exploration 1 Worksheet

The Nose Knows

1. Ms. Foushen needs your help sniffing out the solution to a problem. What has she asked you to do?

Ms. Foushen wants to know which of five smelly chemicals would be the best choice for use in a fire alarm at her school for hearing- and sight-impaired students.

(Recommended 10 pts.)

2. Explain the process of diffusion. (Hint: Check out the wall chart in the lab.)

Sample answer: Diffusion is caused by the constant motion of the particles of matter. It is the process in which particles of a substance move from an area of higher concentration to areas of lower concentration until the particles reach a uniform concentration.

(Recommended 20 pts.)

3. What is the difference between diffusion and osmosis? (If you're not sure, check out the CD-ROM articles.)

Sample answer: Osmosis is diffusion through a semipermeable membrane. The membrane permits the passage of some types of particles while preventing the passage of other types. Diffusion takes place among all particles without a semipermeable membrane.

(Recommended 10 pts.)

4. What is the equipment on the front table in Dr. Labcoat's lab designed to do?

Sample answer: It is designed to show how quickly the particles of each smelly chemical diffuse among air particles.

(Recommended 10 pts.)

Exploration 2 Worksheet

Sea the Light

1. Ms. Sittie wants to create an underwater hanging lamp. What help does she need from you? *(Recommended 10 pts.)*

   Ms. Sittie needs to know which ballast disk to add to a waterproof lamp base to make the entire lamp neutrally buoyant underwater where she dives.

2. What does *ballast* mean? (If you aren't sure, use the CD-ROM articles to help you.) *(Recommended 5 pts.)*

   The term *ballast* refers to something heavy that is used to add mass to an object such as a ship.

3. What purpose do you think the ballast disks serve in the design of the underwater lamp? *(Recommended 5 pts.)*

   Depending on which ballast disk is used in combination with the lamp base, the underwater lamp will rise, sink, or be neutrally buoyant underwater.

4. Describe how you will use the equipment on the lab's front table to answer Ms. Sittie's questions. *(Recommended 10 pts.)*

   The equipment can be used to measure the volume and the mass of each of the ballast disks so that the density of each disk can be calculated. Finding out the total density of the lamp base with each individual ballast disk will determine which disk Ms. Sittie should use for her hanging lamp.

Exploration 1 Worksheet

FAX

To:	Ms. Dee Foushen (FAX 512-555-7003)
From:	
Date:	
Subject:	The Nose Knows

Which of the five samples do you recommend that I use for the fire alarm?

Alcohol	Cinnamon	Garlic	Perfume	Rotten eggs
	■			

(Recommended scoring: 50 points for cinnamon; 30 points for garlic or perfume; and 0 points for alcohol or rotten eggs.)

Please explain why you chose this sample.

The sample that diffused the fastest, the rotten eggs, is highly toxic. The sample that diffused the second fastest, alcohol, is dangerous to use as a fire alarm because it is flammable. Cinnamon is the sample that diffused the third fastest, and it is nontoxic. Therefore, cinnamon would be the best choice to warn the students of fire the most quickly and safely. *(Recommended 25 pts.)*

Explain how odors spread through a room.

Odors spread through a room by a process called diffusion, during which molecules of one substance move from an area of high concentration to an area of lower concentration until a relatively uniform concentration has been reached. *(Recommended 25 pts.)*

5. Use the equipment to conduct all of the necessary tests, and record your data in the first two columns of the table below. Then use your results to calculate the values for the third column.

Ballast disk	Mass (g)	Volume (mL)	Density (g/mL)
A Copper	1344.0	150	8.96
B Aluminum	1469	550	2.67
C Brass	4280.0	500	8.56
D Titanium	1135.0	250	4.54
E Platinum	5363.0	250	21.45
F Zinc	998.0	140	7.13

(Recommended 10 pts.)

6. Calculate the total density of the entire lamp for each individual ballast disk. (If you aren't sure how to calculate the density of an object with multiple parts, examine the CD-ROM articles.)

Ballast disk	Density
A Copper	(659 g + 1344.0 g) ÷ (1500 mL + 150 mL) = 1.214 g/mL
B Aluminum	(659 g + 1469.0 g) ÷ (1500 mL + 550 mL) = 1.038 g/mL
C Brass	(659 g + 4280.0 g) ÷ (1500 mL + 500 mL) = 2.469 g/mL
D Titanium	(659 g + 1135.0 g) ÷ (1500 mL + 250 mL) = 1.025 g/mL
E Platinum	(659 g + 5363.0 g) ÷ (1500 mL + 250 mL) = 3.441 g/mL
F Zinc	(659 g + 998.0 g) ÷ (1500 mL + 140 mL) = 1.010 g/mL

(Recommended 15 pts.)

7. Examine the materials on the back counter of the lab. Use what you see to explain why the different ballast disks have different densities.

The atoms of different elements have different particle arrangements and different densities. Because density is a physical property of an element, each ballast disk has a different density because each disk is made of different elements.

(Recommended 15 pts.)

8. What is buoyant force? (If you're not sure, check out the CD-ROM articles.)

Buoyant force is the upward force exerted by a fluid on a submerged object. The buoyant force is equal to the weight of the fluid displaced by the object. *(Recommended 15 pts.)*

9. Describe the differences among underwater objects that are positively, negatively, and neutrally buoyant. (Hint: Check out the CD-ROM articles.)

A positively buoyant object will tend to rise underwater because it is less dense than the surrounding water. A negatively buoyant object will sink because its density is greater than the density of the surrounding water. A neutrally buoyant object will neither sink nor rise because its density equals the density of the water. *(Recommended 15 pts.)*

Record your answers in the fax to Ms. Sittie.

Exploration 3 Worksheet

Stranger Than Friction

1. Mr. Cline is hard at work on his design for a new amusement park ride. What information is he seeking from you? *(Recommended 10 pts.)*

   Mr. Cline wants to know which material he should use to construct a slide and to use for the bottom of the toboggans for the Camelback Super Slide. He also wants to know what size to make the toboggans.

2. Describe the equipment Dr. Labcoat has set up on the front lab table and the back counter. *(Recommended 10 pts.)*

   On the front table is a model of the Camelback Super Slide, three different-sized toboggans, and a test figure. On the back counter are four blocks, two with a mass of 50 g and two with a mass of 100 g.

3. What is frictional force, and how does it affect a moving object? (Hint: If you're not sure, check out the CD-ROM articles.) *(Recommended 10 pts.)*

   Frictional force is the force that opposes motion between two surfaces that are touching. Frictional force works to keep an object from moving or slows an object down that is already in motion. The magnitude of frictional force depends on the coefficient of friction and on the normal force.

4. What does the wall chart in the lab show about normal force and the coefficient of friction? *(Recommended 20 pts.)*

   The diagram shows how a greater normal force (the force pressing two objects together) can result in a greater frictional force. Therefore, a greater amount of force is necessary to move heavier objects. It also shows how changing the coefficient of friction between two surfaces can reduce friction and the effects of frictional force.

Exploration 2 Fax Form

FAX

To: Ms. Diane Sittie (FAX 817-555-4459)
From:
Date:
Subject: Sea the Light

Please complete the following chart:

METAL	MASS	VOLUME	DENSITY
Aluminum	1469.0 g	550 mL	2.67 g/mL
Brass	4280.0 g	500 mL	8.56 g/mL
Copper	1344.0 g	150 mL	8.96 g/mL
Platinum	5363.0 g	250 mL	21.45 g/mL
Titanium	1135.0 g	250 mL	4.54 g/mL
Zinc	998.0 g	140 mL	7.13 g/mL

(Recommended 10 pts.)

What is the density of the waterproof lamp base? *(Recommended 10 pts.)*

659 g ÷ 1500 mL = 0.439 g/mL

Please indicate your metal selection for the ballast disk here: **Titanium**

(Recommended scoring: 50 points for titanium; 30 points for aluminum or zinc; and 0 points for brass, copper, or platinum.)

Why did you pick this metal? *(Recommended 30 pts.)*

Sample answer: The total density of the lamp base and the titanium ballast disk is approximately the same as the density of the sea water where Ms. Sittie dives. As a result, the entire lamp will hang in the water at the depth that Ms. Sittie needs. If the lamp base plus the ballast disk were denser than the sea water, then the lamp would sink. If the lamp base plus the ballast disk were less dense than the sea water, then the lamp would float to the surface of the water.

5. Use the force meter on the back lab counter to find the force required to pull each block. Record your results below.

It takes 0.25 N of force to pull both 50 g blocks at the same speed. It takes 0.50 N of force

to pull both 100 g blocks at the same speed.   *(Recommended 15 pts.)*

6. Does the amount of surface area touching the block affect the force required to pull it? Why or why not?

Both 50 g blocks require the same amount of force to move across the counter at the same speed, even though one block is on its side and the other is on its edge. The same is true for both 100 g blocks. This shows that the force of friction is not determined by the area of contact between two surfaces.   *(Recommended 15 pts.)*

7. Conduct the necessary tests with the prototype for the Camelback Super Slide, and record your results in the table below. (Hint: It may not be necessary to try every possible combination.)

Material component for slide	Material component for toboggan	Toboggan size (cm)	Observations
teflon	teflon	any	Toboggan loses contact with the slide at the hump. Figure flies off the slide.
teflon	stainless steel	any	Toboggan slips on the hump and ejects figure. Both land at bottom of slide.
teflon	plastic	any	Toboggan gets stuck and slides back and forth in dip before hump.
stainless steel	teflon	any	Toboggan slips on the hump and ejects figure. Both land at bottom of slide.
stainless steel	stainless steel	any	Toboggan stays in contact with slide all the way down, giving the best ride.
stainless steel	plastic	any	Toboggan gets stuck and slides back and forth in dip before hump.
plastic	teflon	any	Toboggan gets stuck and slides back and forth in dip before hump.
plastic	stainless steel	any	Toboggan gets stuck and slides back and forth in dip before hump.
plastic	plastic	any	Toboggan gets stuck and slides back and forth in dip before hump.

(Recommended 20 pts.)

Record your answers in the fax to Mr. Cline.

Exploration 3
Fax Form

FAX

To: Mr. Norm N. Cline (FAX 281-555-5276)

From:

Date:

Subject: Stranger Than Friction

What material do you recommend for the construction of the slide?

Stainless steel

What material do you recommend for the construction of the toboggan?

Stainless steel

What is your recommendation regarding toboggan size?

100 cm
120 cm
140 cm
any of the above

(Recommended scoring: 50 points for stainless steel for both the slide and the toboggans and "any of the above" selected; 30 points for one or two incorrect materials chosen and "any of the above" selected; 30 points for stainless steel for both the slide and the toboggans and "any of the above" not selected; 0 points for one or two incorrect materials and "any of the above" not selected.)

What effect does the size of the toboggan have on the performance of the Camelback Super Slide? Explain.

The size of the toboggan has no effect on the performance of the ride. The friction between two given surfaces is the same for a given normal force no matter how much area is in contact between the two surfaces.   *(Recommended 50 pts.)*

Latitude Attitude

Exploration 4
Worksheet

1. Capt. Corey O. Lease needs to get supplies to a group of scientists on the Mertz Glacier. What has she asked you to do for her?

   Capt. Lease has asked Dr. Labcoat to verify that her flight direction and average speed calculations are appropriate for a trip from Melbourne to the Mertz Glacier.

   (Recommended 10 pts.)

2. In relation to Melbourne, where is the Mertz Glacier located?

   The Mertz Glacier is due south of Melbourne. *(Recommended 10 pts.)*

3. How does the speed of the Earth's surface at the equator compare with the speed of the Earth's surface at the poles? (If you're not sure, check out the CD-ROM articles.)

   Points on the Earth's surface at the equator have greater velocities (about 1670 km/hr) than do points at the poles (effectively 0 km/hr).

   (Recommended 15 pts.)

4. How will you use the Accu-Flight Simulator to help Capt. Lease?

   Sample answer: I will use the Accu-Flight Simulator to examine the results of different combinations of flight direction and airspeed. After selecting a certain flight direction and airspeed, I will look at the display screen to see where Capt. Lease's plane would end up. When I find the combination of flight direction and airspeed that takes the plane directly to the Mertz Glacier while maintaining fuel efficiency, I will recommend this flight plan to Capt. Lease.

   (Recommended 15 pts.)

5. Use the Accu-Flight Simulator to determine the results of using different combinations of flight direction and average airspeed. Record your results in the table below.

Flight direction	Airspeed (km/hr)	Final location of the plane
130°	725	**east of the Mertz Glacier, off the screen**
140°	725	**east of the Mertz Glacier, off the screen**
150°	725	**northeast of the Mertz Glacier, in the ocean**
180°	725	**northeast of the Mertz Glacier, in the ocean**
210°	725	**on Antarctica, but east of the Mertz Glacier**
220°	725	**on Antarctica, slightly east of the Mertz Glacier**
230°	725	**on Mertz Glacier**
130°	850	**east of the Mertz Glacier, off the screen**
140°	850	**east of the Mertz Glacier, off the screen**
150°	850	**northeast of the Mertz Glacier, in the ocean**
180°	850	**on Antarctica, but east of the Mertz Glacier**
210°	850	**on Antarctica, but east of the Mertz Glacier**
220°	850	**on Mertz Glacier**
230°	850	**on Antarctica, slightly west of the Mertz Glacier**

(Recommended 15 pts.)

6. What does the equipment on the lab's back counter demonstrate?

   The equipment demonstrates why the Coriolis effect occurs. The turntable rotates clockwise about one-third of a revolution. As a result, the pencil line curves to the left. This is similar to what happens to Capt. Lease's plane as it travels toward Antarctica.

   (Recommended 15 pts.)

7. Describe how moving objects are affected by the Coriolis effect. (If you're not sure, check out the CD-ROM articles.)

   The Coriolis effect influences the paths of moving objects such as airplanes, winds, and weather systems. In the Northern Hemisphere, objects moving north or south seem to veer to their right (clockwise). In the Southern Hemisphere, the curvature of the path of objects moving north or south is to their left (counterclockwise).

   (Recommended 20 pts.)

Record your answers in the fax to Capt. Lease.

Exploration 5 Worksheet

Tunnel Vision

1. Seymore Rhodes wants to make his mountain-bike ride to school a safer one. What advice does he need from you?

   Seymore wants to know how to brighten the lights in a circuit for his bicycle helmet. He also wants to make sure that the circuit is lightweight.   *(Recommended 10 pts.)*

2. What are the necessary components of an electric circuit? (If you're not sure, check out the CD-ROM articles.)

   The components of an electric circuit include a source of electricity, such as a battery; an output device, such as a light bulb; and wires. Many circuits also include switches, which allow the circuit to be turned on and off.   *(Recommended 10 pts.)*

3. Look at the diagram that Seymore drew of his circuit. What kind of circuit is it?

   Seymore's circuit is a series circuit.   *(Recommended 5 pts.)*

4. How is a series circuit different from a parallel circuit? (Hint: Check out the equipment on the lab's back counter.)

   In a series circuit, there is only one pathway for the current to follow. If there is a break in the circuit, the flow of current is disrupted and none of the bulbs light. In a parallel circuit, current is divided into two or more branches. If a break occurs in one branch, current can continue to flow through other branches and the bulbs along those branches stay lit. On the back counter, the bulbs in the series circuit are not as bright as those in the parallel circuit. This is because the bulbs connected in series share the same current, while each light bulb connected in parallel receives the full amount of current provided by the battery.

   (Recommended 15 pts.)

Exploration 4 Fax Form

FAX

To: Capt. Corey O. Lease (FAX 011-612-555-7996)

From:

Date:

Subject: Latitude Attitude

In which compass direction should I aim my plane?

180° (due south)	130°	140°	150°	210°	220°	230°
						■

How fast should I fly the plane?

725 km/hr	850 km/hr

(Recommended scoring: 50 points for 230° and 725 km/hr; 35 points for 220° and 850 km/hr; and 0 points for anything else.)

Please explain why it is necessary for me to fly the plane in the recommended direction in order to reach the Mertz Glacier.

Sample answer: Points on the Earth's surface at higher latitudes have less surface velocity than those at latitudes closer to the equator. As a result, the paths of objects that move above the Earth's surface appear curved. This phenomenon is called the Coriolis effect. For example, an object moving south in the Southern Hemisphere, such as Capt. Lease's plane, is deflected counterclockwise, or east, in relation to the ground. Because of the Coriolis effect, Capt. Lease must aim her plane slightly southwest to reach her destination. The exact direction in which she must fly depends on how fast the plane is traveling. For example, if Capt. Lease flies her plane at an average speed of 850 km/hr, the angle southwest at which she flies the plane may be less than it would be if she flies her plane at 725 km/hr.

(Recommended 50 pts.)

7. What effect does the number of batteries have on your circuits?
   The number of batteries affects the brightness of the bulbs because the batteries determine the amount of current that can flow through the circuits created in this Exploration. For example, in a series circuit consisting of three bulbs, adding a second battery made the bulbs brighter. In a parallel circuit consisting of three bulbs, adding a second battery caused the bulbs to burn out. Adding a second battery also increases the total weight of the circuit.
   (Recommended 10 pts.)

8. What happens when too much current flows through a circuit?
   When too much current flows through a circuit, the circuit can fail. For example, if too much current is introduced into a circuit involving a light bulb, the bulb can burn out.
   (Recommended 10 pts.)

9. What is the relationship between current, voltage, and resistance? (If you're not sure, check out the CD-ROM articles.)
   Ohm's law describes the relationship between current, voltage, and resistance. For a given resistance, current and voltage are directly proportional according to the equation $I = E/R$. That means that, for a given resistance, increasing the voltage increases the current.
   (Recommended 10 pts.)

Record your answers in the fax to Seymore Rhodes.

---

5. Examine the equipment on the lab's front table. How will you use the switches and the batteries to help you answer Seymore's questions?
   Sample answer: I will use the switches to create various circuits. By closing certain switches, I can observe how different circuits affect the operation and brightness of the three light bulbs. I will use the batteries to determine how greater amounts of voltage affect the brightness of the bulbs.
   (Recommended 10 pts.)

6. Create all of the necessary circuits you need to answer Seymore's questions. In the table below, record the switches you close, the number of batteries you use, and the type of circuit you create. Use the fourth column to record your observations about the light bulbs.
   Please note: The following are sample results. Many more combinations are possible.

Switches closed	Number of batteries	Type of circuit	Effect on light bulbs
1, 3, 6	1	series	A and B are somewhat bright.
1, 2, 4, 6	1	parallel	A and B are very bright.
1, 3, 5, 8	1	series	A, B, and C are dim. (This is Seymore's circuit.)
1, 3, 6, 7, 8	1	series-parallel	A and B are somewhat bright; C is very bright.
1, 2, 4, 5, 8	1	series-parallel	A is very bright; B and C are somewhat bright.
1, 2, 4, 6, 7, 8	1	parallel	A, B, and C are very bright. (This is the best circuit.)
1, 3, 5, 8	2	series	A, B, and C are somewhat bright.
1, 2, 4, 6	2	parallel	A and B burn out.
1, 2, 7, 8	2	parallel	A and C burn out.
1, 2, 4, 5, 8	2	series-parallel	A burns out; B and C are very bright.
1, 2, 4, 6, 7, 8	2	parallel	A, B, and C burn out.

(Recommended 20 pts.)

Describe the changes that you made to Seymore's original circuit.

Seymore's original circuit is a series circuit. Because the three light bulbs are connected in series, they share the same current. As a result, the current produced by the single battery is only strong enough to dimly light all three bulbs. In order to improve Seymore's circuit, it is necessary to connect the bulbs in parallel. This means that each light bulb will receive the full strength of the current provided by the battery, and the bulbs will be very bright. *(Recommended 25 pts.)*

Exploration 5
Fax Form

FAX

To: Seymore Rhodes (FAX 406-555-7209)

From:

Date:

Subject: Tunnel Vision

How many batteries should I use? **1** **2**

Please describe the best way to wire the lights for my helmet.

The best way to wire the lights for the bicycle helmet is in parallel with one battery.
With this combination, the bulbs are very bright and the circuit is still lightweight.
(Recommended 25 pts.)

For Internal Use Only

Please answer the following questions for my laboratory records. Scientists must always keep good records. Dr. Crystal Chroma

What is the best type of circuit for Seymore's lights?

| SERIES | SERIES-PARALLEL | **PARALLEL** |

(Recommended scoring: 50 points for one battery and parallel; 30 points for two batteries and series; and 0 points for any other answers.)

Exploration 6 Worksheet

Sound Bite!

1. Mr. Lintz is worried about the aggressive behavior of his guinea pigs. Describe his problem and what he has asked you to do.

   The guinea pigs in Mr. Lintz's pet store have been biting the customers. Mr. Lintz suspects that the loud humming noise from the newly installed freezers next door is driving his guinea pigs to violence. He wants to know how to use active noise control to eliminate this noise pollution.

   (Recommended 10 pts.)

2. Examine the equipment on the back counter of the lab, and then describe what sound waves are.

   As demonstrated by the equipment, air particles spread away from a vibrating sound source, such as the drum, and push against air particles in front of them. The particles continue to crowd together and spread apart in succession as the sound travels away from its source, and the compressions and rarefactions of longitudinal waves of sound result.

   (Recommended 15 pts.)

3. What purpose does an oscilloscope serve? (Hint: Check out the wall chart.)

   An oscilloscope is an instrument that converts sound waves (longitudinal waves) to transverse waves that appear on a screen. This allows us to "see" sound waves.

   (Recommended 10 pts.)

4. What is destructive interference? (If you're not sure, check out the CD-ROM articles.)

   Destructive interference occurs when the crest of one wave meets the trough of another wave. This causes a decrease in amplitude. If a crest and trough of equal amplitude and frequency meet, they cancel each other out.

   (Recommended 15 pts.)

EXPLORATION 6 • SOUND BITE! 199

---

5. Conduct your experiment using the lab equipment. In the table below, record the frequency, amplitude, and phase you selected for the generated sound. Use the fourth column to describe the wave you see in the center oscilloscope screen when the generated sound is combined with the offending sound.

   Please note: The following answers are sample observations. Many more combinations are possible.

Frequency (Hz)	Amplitude	Phase	Results
175	low	1	A wave with varying amplitude. When the amplitude is at its greatest, it is greater than the amplitude of both the offending sound and the generated sound.
225	low	1	A wave with constant amplitude. The amplitude is twice that of either the offending sound or the generated sound.
275	low	1	A wave with varying amplitude. When the amplitude is at its greatest, it is greater than the amplitude of both the offending sound and the generated sound.
325	low	1	A wave with varying amplitude. When the amplitude is at its greatest, it is greater than the amplitude of both the offending sound and the generated sound.
175	med.	1	A wave with varying amplitude. When the amplitude is at its greatest, it is greater than the amplitude of both the offending sound and the generated sound.
225	med.	1	A wave with constant amplitude. The amplitude is more than twice that of the offending sound.
325	high	1	A wave with varying amplitude. When the amplitude is at its greatest, it is greater than the amplitude of both the offending sound and the generated sound.
175	low	2	A wave with varying amplitude. When the amplitude is at its greatest, it is greater than the amplitude of both the offending sound and the generated sound.
225	low	2	A flat line appears on the screen. The crests and troughs of the waves of the offending sound and the generated sound cancel out.
275	med.	2	A wave with varying amplitude. When the amplitude is at its greatest, it is greater than the amplitude of both the offending sound and the generated sound.
225	high	2	A wave with constant amplitude. The amplitude is more than twice that of the offending sound but less than the generated sound.

(Recommended 25 pts.)

Exploration 6
Fax Form

FAX

To: Mr. Cy Lintz (FAX 707-555-8988)

From:

Date:

Subject: Sound Bite!

Which settings eliminate Mr. Lintz's noise pollution?

Frequency (hertz)	175	225	275	325
Amplitude	low	medium	high	
Phase	1	2		

(Recommended scoring: 50 points for all three correct answers; 30 points for one or two correct answers; and 0 points for no correct answers.)

For Internal Use Only

Please answer the following questions for my laboratory records. Scientists must always keep good records. *Dr. Crystalofsweat*

Explain how sound travels.

Sample answer: Sound waves are longitudinal waves that travel through a medium, such as air. As a source of sound vibrates, it alternately pushes the air molecules together, forming compressions, and spreads them apart, forming rarefactions. A series of compressions and rarefactions form and spread out from the source of the sound.

(Recommended 25 pts.)

202   HOLT SCIENCE AND TECHNOLOGY INTERACTIVE EXPLORATIONS TEACHER'S GUIDE

---

6. Based on your results, what are the offending sound's frequency, amplitude, and phase?
The offending sound is in phase 1 and has a frequency of 225 Hz and a low amplitude.
(Recommended 10 pts.)

7. What does *phase* refer to? (If you're not sure, check out the CD-ROM articles.)
***Phase* is a term used to describe the positions of the crests and troughs of two individual waves relative to each other.**
(Recommended 15 pts.)

Record your answers in the fax to Mr. Lintz.

EXPLORATION 6 • SOUND BITE!   201

ANSWER KEY • DISC 3

DISC 3 • ANSWER KEY   273

Exploration 7 Worksheet

In the Spotlight

1. Ms. Kones is a little in the dark about the lighting design for her first stage production. What questions do you need to answer for her?

   Sample answer: **Ms. Kones wants to know how many different colors of light she can create using her filters, what those colors are, and what filter combinations are required to make those colors. She also wants to know why blue light added to yellow light results in white light.**

   (Recommended 10 pts.)

2. Describe the equipment that Dr. Labcoat has set up on the front table.

   There is a white screen, two spotlights, and 7 filter selections (including no filter) installed in each spotlight.

   (Recommended 5 pts.)

3. How do colored filters work to create different colors of light? (If you aren't sure, check out the CD-ROM articles.)

   A filter is a colored lens that absorbs certain wavelengths of light and transmits others. When white light passes through a filter, the filter absorbs all the colors of light except the color of light that matches the color of the filter. A red filter transmits only red light and absorbs every other color of light. Because the filter absorbs some colors of light, using filters is a kind of subtractive color formation. Filters can also be used in additive color formation. By mixing two different colors of filtered light, new colors can be produced.

   (Recommended 15 pts.)

Explain how active noise control is used to eliminate noise pollution.

Sample answer: **Active noise control uses destructive interference to reduce a sound's amplitude. A sound of the same frequency as the noise pollution is used to cancel out the offending noise. In some cases, a speaker is set up to produce sound waves that are the mirror image of the noise pollution. When the sound waves from the speaker meet the sound waves from the source of noise, the waves cancel each other out.**

(Recommended 25 pts.)

4. As you try different combinations of filters, record your results, including the color equations for each color, in the table below.

Filter 1	Filter 2	Resulting color	Color equation
no filter	no filter	white	no filter + no filter = white
no filter	red	red	no filter + red = red
no filter	cyan	cyan	no filter + cyan = cyan
no filter	yellow	yellow	no filter + yellow = yellow
no filter	blue	blue	no filter + blue = blue
no filter	green	green	no filter + green = green
no filter	magenta	magenta	no filter + magenta = magenta
red	red	red	red + red = red
red	cyan	white	red + cyan = white
red	yellow	burnt orange	red + yellow = burnt orange
red	blue	magenta	red + blue = magenta
red	green	yellow	red + green = yellow
red	magenta	raspberry	red + magenta = raspberry
cyan	cyan	cyan	cyan + cyan = cyan
cyan	yellow	green	cyan + yellow = green
cyan	blue	medium blue	cyan + blue = medium blue
cyan	green	pale green	cyan + green = pale green
cyan	magenta	blue	cyan + magenta = blue
yellow	yellow	yellow	yellow + yellow = yellow
yellow	blue	white	yellow + blue = white
yellow	green	lemon lime	yellow + green = lemon lime
yellow	magenta	red	yellow + magenta = red
blue	blue	blue	blue + blue = blue
blue	green	cyan	blue + green = cyan
blue	magenta	vivid violet	blue + magenta = vivid violet
green	green	green	green + green = green
green	magenta	white	green + magenta = white

(Recommended 20 pts.)

5. What are the primary colors of light?
   The primary colors of light are red, blue, and green. *(Recommended 10 pts.)*

6. What are the secondary colors of light?
   The secondary colors of light are yellow, cyan, and magenta. *(Recommended 10 pts.)*

7. What happens when you mix two primary colors of light?
   Mixing two primary colors of light results in a secondary color of light. *(Recommended 10 pts.)*

8. What happens when you mix two secondary colors of light?
   Mixing two secondary colors of light results in a primary color of light. *(Recommended 10 pts.)*

9. What differences do you notice between mixing the colors of paint on the back counter and mixing colors of light? (If you aren't sure, check out the CD-ROM articles.)
   Sample answer: When you mix colors of paint, more colors are absorbed than are reflected, so the combinations move toward "blackness" instead of "whiteness." As a result, red + blue + yellow = "black." Mixing colors of paint is subtractive color formation, unlike mixing colors of light, which is additive color formation. *(Recommended 10 pts.)*

Record your answers in the fax to Ms. Kones.

Describe how colored filters produce different colors of light.

Sample answer: A filter is a colored lens that absorbs certain wavelengths of light and transmits others. For example, when white light is passed through a red filter, red light is transmitted and all other colors of light are absorbed. When two or more beams of filtered light overlap on a white screen, they combine to form a new color of light.

(Recommended 15 pts.)

Please explain the difference between mixing colors of paint and mixing colors of light.

Sample answer: Mixing colors of paint is called subtractive color formation because it tends to subtract colors of light from white light. For example, if you mix two colors of paint together, the mixture absorbs more colors of light than either paint would absorb separately. When you mix various colors of light, you are adding different wavelengths of light together and "building" white light. This is called additive color formation.

(Recommended 20 pts.)

Exploration 7
Fax Form

FAX

To: Ms. Iris Kones (FAX 409-555-2017)

From:

Date:

Subject: In the Spotlight

How many different colors of light (including white light) can be produced using Ms. Kones's two spotlights and twelve filters? __**13**__ *(Recommended scoring: 50 points for 13 colors; 30 points for 10, 11, or 12 colors; and 0 points for 9 or fewer colors.)*

Write the color equation for each color that you produced.

White: no filter + no filter = white; yellow + blue = white; green + magenta = white; red + cyan = white

Red: red + no filter = red; red + red = red; yellow + magenta = red

Green: green + no filter = green; green + green = green; cyan + yellow = green

Blue: blue + no filter = blue; blue + blue = blue; cyan + magenta = blue

Magenta: magenta + no filter = magenta; magenta + magenta = magenta; red + blue = magenta

Yellow: yellow + no filter = yellow; yellow + yellow = yellow; red + green = yellow

Cyan: cyan + no filter = cyan; cyan + cyan = cyan; blue + green = cyan

Burnt Orange: red + yellow = burnt orange

Raspberry: red + magenta = raspberry

Medium Blue: cyan + blue = medium blue

Pale Green: cyan + green = pale green

Lemon Lime: yellow + green = lemon lime

Vivid Violet: blue + magenta = vivid violet

(Recommended 15 pts.)

Exploration 8 Worksheet

DNA Pawprints

1. Ms. Jean Poole wants to enter her dogs in an upcoming dog show. What does she need to know in order to complete the pedigrees for her dogs?

   Ms. Poole needs to know which of her older male dogs fathered each of her three younger dogs. She also wants to know a little bit about the test used to determine who sired which dog.

   (Recommended 5 pts.)

2. What does DNA have to do with inherited characteristics? (If you're not sure, check out the CD-ROM articles.)

   Sample answer: DNA is the chemical responsible for the makeup of our genes. DNA is found in almost every one of our cells, and it acts as a set of instructions for how we look. We inherit our genes from our parents, who inherited their genes from their parents. The millions of arrangements of nucleotides in strands of DNA account for the vast differences in our individual genetic makeup.

   (Recommended 10 pts.)

3. Explain how DNA fingerprinting works. (Hint: Check out the wall chart.)

   DNA fingerprinting is a process that allows you to compare different samples of DNA. An electric current sorts fragments of DNA by size in a gel tray. The fragments are transferred to a thin film that is bathed in a radioactive solution. The radioactive bath marks certain pieces of DNA, which then show up as bands in an X ray.

   (Recommended 10 pts.)

4. Describe the setup on the front table in the lab.

   There are DNA samples from Bella, Domino, Merlin, Sugar, Duke, King, and Roy; sterile tips for a micropipet; a gel tray; an electrophoresis chamber; a gel processor; and an X-ray developer.

   (Recommended 10 pts.)

5. Conduct DNA fingerprinting for each young dog, mother, and possible father, and record your results in the table below.

Mother	Young dog	Possible father	Observations of DNA fingerprints
Bella	Domino	Duke	None of Domino's bands match Duke's bands. Two of Domino's bands match Bella's bands.
Bella	Domino	King	None of Domino's bands match King's bands. Two of Domino's bands match Bella's bands.
Bella	Domino	Roy	Two of Domino's bands match Roy's bands. Two of Domino's bands match Bella's bands.
Bella	Merlin	Duke	None of Merlin's bands match Duke's bands. Two of Merlin's bands match Bella's bands.
Bella	Merlin	King	Two of Merlin's bands match King's bands. Two of Merlin's bands match Bella's bands.
Bella	Merlin	Roy	None of Merlin's bands match Roy's bands. Two of Merlin's bands match Bella's bands.
Bella	Sugar	Duke	None of Sugar's bands match Duke's bands. Two of Sugar's bands match Bella's bands.
Bella	Sugar	King	Two of Sugar's bands match King's bands. Two of Sugar's bands match Bella's bands.
Bella	Sugar	Roy	None of Sugar's bands match Roy's bands. Two of Sugar's bands match Bella's bands.

(Recommended 30 pts.)

6. How does Bella's DNA fingerprint compare with the DNA fingerprints of Domino, Merlin, and Sugar?

Because Bella is the mother of the three dogs, each young dog gets half of its DNA from Bella. As a result, two bands on each young dog's DNA fingerprint will match two bands on Bella's DNA fingerprint.

(Recommended 15 pts.)

7. How can you tell which older male fathered each young dog?

In order for an older male and a younger male dog to be related, they must have some of the same DNA. Half the young dog's DNA must come from the mother, and the other half must come from the father. If half the bands of the DNA fingerprints of a young dog and an older male dog match, then they are related.

(Recommended 15 pts.)

8. Look at the material on the lab's back counter. What does it tell you about where DNA is located?

DNA is found in the nucleus of almost every cell in all living organisms.

(Recommended 5 pts.)

Record your answers in the fax to Ms. Poole.

Exploration 8
Fax Form

FAX

To: Ms. Jean Poole (FAX 512-555-8163)

From:

Date:

Subject: DNA Pawprints

Please indicate which male sired Domino, Merlin, and Sugar.

Domino	○ Duke	○ King	● Roy
Merlin	○ Duke	● King	○ Roy
Sugar	○ Duke	● King	○ Roy

(Recommended scoring: 50 points for all three correct matches; 30 points for one or two correct matches; and 0 points for no correct matches.)

Describe the test that you used to determine which sire matched each of the young dogs.

The test used to determine which sire matched each of the young dogs is called DNA fingerprinting. DNA samples from each dog were sliced into fragments and separated according to size with an electric current. The samples were then marked with a radioactive tag, and an X ray was taken of them to reveal the DNA fingerprint. Each young dog's DNA fingerprint could then be compared with the DNA fingerprints of Bella and each possible father to determine the heredity of each young dog.

(Recommended 30 pts.)